NORTON SINGLES

OSPREY
COLLECTOR'S
LIBRARY

NORTON SINGLES

Manx and Inter, and all side and overhead valve singles from 1927 to 1966

Roy Bacon

First published in 1983 by Osprey Publishing Limited
27A Floral Street, London WC2E 9DP
Member company of the George Philip Group
First reprint winter 1985
Second reprint spring 1986

British Library Cataloguing in Publication Data
Bacon, Roy
Norton singles. (Osprey collector's library)
1. Norton motorcycle
I. Title
629.2'275 TL448.N6
ISBN 0-85045-485-9

Editor Tim Parker
Design Roger Daniels
Filmset and printed in England by
BAS Printers Limited, Over Wallop, Hampshire

Contents

Geoff Duke at Oulton Park on his special with Reynold frame and leading link forks. Lovely style

Foreword by Geoff Duke

I joined Norton's in January, 1948 and very proud I was too of the opportunity to be part of such an illustrious organisation! Although, of course, in the beginning I was just a trials rider; that other 'pinnacle' took two more years to achieve. In those days, every budding young road racer's ambition was to be invited to ride for Norton.

I read with great interest, and not a little nostalgia, a copy of the manuscript of *Norton Singles*, and I was truly amazed by the accuracy of the mass of detailed information covering just *my* five year's association with the marque, and no doubt this is reflected even in the earliest days.

The enthusiasm of the work force at Bracebridge Street had to be seen to be believed. Every Monday morning, if the previous weekend's race happened to be in this country, I would be inundated with, 'How did you get on?' as I walked across the factory floor. Then back to the production line they went, usually with renewed enthusiasm, which must have paid dividends to the shareholders!

The great names that crop up from time to time in *Norton Singles*, Alec Bennett, Jimmy Simpson—my personal boyhood hero—Harold Daniell, and not forgetting the remarkable Joe Craig, all long since gone but never to be forgotten characters, each one of them a master of his art.

This book doesn't seek to hide the many trials and tribulations inevitably experienced in such a long history, so I must recount just one of the many which nevertheless had its amusing side.

At the Dutch TT, the Norton team was based at a small hotel at the end of the long Hooghalen 'straight', and it was there that Joe Craig checked the 'full-bore' carburettor settings of each machine by taking out the spark plug and checking its colour. This involved the rider in cutting his engine and freeing the clutch to coast the last few yards to where Joe and the mechanics were stationed. In the jargon of those days this was known as a 'chop-out', and on this particular occasion, Harold Daniell free-wheeled to a standstill alongside Joe and announced that he had just achieved the perfect chop-out. Joe's eyes followed Harold's pointing finger to where a large hole in the crankcase of the engine revealed the jagged end of a broken connecting rod! The technical director was not amused.

Famous makes may come and go, but Norton lives on forever. Mention the marque now, to any reasonably knowledgeable follower of road racing, and invariably, with a sigh, just two words are uttered, 'Ah, Norton!'.

Geoff Duke OBE
Six times world champion
Douglas, Isle of Man
November 1982

Acknowledgements

I learnt about the forgiving nature of the Unapproachable Norton early one morning when my hard old tyres let go when flat-in-third halfway round a left curve on the Robin Hood Way and my 1936 ES2 slid right across the road. Some wheelgrip was found a foot from the right kerbstone, the old Norton shook its head and continued on with its young, stupid and very white rider wondering why he was not in the ditch. It was Norton handling that saved a spill that day, and others, and that machine set my standards for roadholding, braking and gear-changing. It was some time before I rode anything else as good.

Later I had an ex-WD 16H and float to transport my racing Bantam so this book brought back the memories. That side valve machine did sterling work and sadly went when a car turned it over while a friend was transporting his elderly KTT.

In writing this book I owe a great debt to Titch Allen and his Norton Saga published in *Motor-cycle Sport* around 1970, together with his later articles in that magazine recounting his wartime experiences. His work made my task very much easier and I am further in his debt for he read and checked my manuscript and appendices, adding his own pithy comments.

I am also indebted to many other people who bore with my telephone calls and answered my letters to provide information and hunt up pictures. These numbered Charlie Juggins who had been the Bracebridge Street paint shop foreman; Don Morley; for data on the 597 ohc machines Mick Woollett, Steve Lancefield Francis Beart, Cyril Quantrill and Denis Jenkinson, especially Jenks for his first hand knowledge of the twin-cam motor used by Oliver; John Wilkinson who owned the plunger Big 4; John Hudson for data on this and machine numbering; Chris Richardson who lent me a Manx brochure and put me in touch with two picture sources in the USA and a *MCN* 'Ask' reader who did the same for me in South Africa.

The pictures and line drawings come mainly from magazine files and I am again indebted to Bob Berry and Peter Law of *Motor Cycle News*, Mick Woollett of *Motor Cycle Weekly*, Jim Lindsay of *Mechanics*, Cyril Ayton of *Motorcycle Sport* and Mike Nicks and Tim Holmes of *Classic Bike*. Pictures were obtained from the Imperial War Museum and the Montagu Motor Museum at Beaulieu. Help from abroad was from Barry Sewell in South Africa and his friend Dennis while in California Paul Adams took shots of his vintage 16H and early Manx at my request and Art Sirota sent a shot of him being cited for no mirror on his lovely 1947 model 18. Nearer home pictures came from Titch Allen, Charles Mortimer and John Wilkinson.

Pictures from the professionals were also used and in the pages are the work of Allsport, T M Badger, Gerald Betteridge, Cunningham-Clayton, Mirco Decet, I M Frost, Island Photographics, K G Jones, T C March, Morland Braithwaite, Don Morley, Nick Nicholls, Tim Parker, Bill Salmond and finally Franco Zagari for a lovely Ray Amm shot.

Finally, I must thank Geoff Duke for kindly writing the foreword and, as always Tim Parker for his never ending assistance.

Roy Bacon
Niton, Isle of Wight
January 1983

1 Formative years

The Norton single is for many the epitome of the marque, representing a continuous pattern of development that ran for nearly half a century. In that timespan the engine design took only a few major steps but each was to establish a new standard, new legends, and its own charisma. All could be seen to come from the same mould laid down by the company's founder at the start with good, honest and upright standards, nothing too clever but well made and durable. And so generation after generation of rider came to know the same magic numbers—Big 4, CS1, 18, ES2, 16H—a sequence that in time ranged from models 1 to 55.

James Lansdowne Norton was born in Birmingham in 1869, and in 1898 founded his Norton Manufacturing Company in that city. He made fixtures and fittings for the cycle and infant motorcyle industries and so came to meet many of the pioneers of the business. One of these was Charles Riley Garrard who imported small clip-on power units for bicycles under the name Clement-Garrard, these coming from M. Clement in France. In 1902 one was fitted to the first machine to bear the Norton name.

This machine with engine mounted on the front downtube was followed by others and an early indication of Norton's high standards and wish to provide the best for his customers was the option of a two-speed countershaft gear and all chain drive. His early machines were assembled from parts made in and around the city but

James Lansdowne Norton on his Big 4 fitted with footboards and with sidecar attached

1911 Norton single with old style tank insignia and classic engine line

it was he who was the driving force that brought the components together and pushed forward with new ideas. In time his premises became an informal business club and a centre for men working in the industry to meet at in the evening and chat about mutual problems.

J. L. Norton was a man of the highest standards, totally honest and deeply religious with a love for motorcycling. He had an early interest in engineering and the steam engines he built at home as a boy attracted such crowds of interested onlookers when he set them running that the police asked him to run them elsewhere to prevent obstruction.

Norton was apprenticed to a toolmaker and learnt his skills on the intricate toolwork used in the jewellery trade. He was an idealist and allowed nothing short of perfection out of the works. In the early days he ensured this by acting as chief tester.

He suffered poor health which brought about his death in 1925 at the age of 56. His appearance, with patriarchal beard and lined face, meant that he and his father were often mistaken for brothers when out together on a sidecar with son at the helm and father as passenger. In spite of his health Norton was to ride in three TTs (1909 to 1911) and attended the very first when Rem Fowler won the multi-cylinder class on his Peugeot-engined Norton to begin the firm's long association with road racing and the Isle of Man.

Due to his appearance, his religion, the high

standards he sought for the industry regardless of their effect on the business, and his forward approach to technical matters, James Norton became known as 'Pa Norton' and later as the 'Father' of the industry. His pleasure was riding and creating good machines for others to ride.

In the Edwardian period while Norton was still using imported French engines he moved his frame design away from the bicycle form to lower the saddle and increase the wheelbase which improved the ride. The engine was mounted well forward and the machines became nicknamed 'ferrets', but it was to become the classic 'vintage' style, with the addition of a gearbox, in time.

In 1907 the years of experience resulted in the appearance of the first Norton single prototype and it set a style and form that was to run on to 1954. Models reached the public in 1908 and the machine was called the Big 4. The first engine had dimensions of 82 × 125 mm giving a capacity of of 660 cc, but in production the stroke was reduced to 120 mm and the capacity to the classic 633 cc. Actually 633·7 cc so 634 would have been more accurate, but 633 was the number to run down the years.

The engine was a simple side valve with a circular aluminium crankcase, cast iron head and barrel in one, valves on the right and magneto mounted in front of the crankcase driven by chain from the exhaust camshaft. Characteristically, the exhaust pipe was nearly vertical and ran between magneto and crankcase to a cylindrical silencer mounted across the machine in front of the engine. Already Norton was very aware of the need to allow the gases an easy exit from the cylinder as well as an unobstructed entrance. He also appreciated that the limitations of the engine revolved around the heat feed back from the exhaust port and hence the need to isolate it from the cylinder and keep it straight with an air space for cooling air. It was

Vic Horsman at Brooklands with his all chain drive machine

Norton
37

these fundamental design points that were to allow the dramatic development of the engine in the 1920s.

The remainder of the machine was straightforward, with belt drive, the long low Norton frame, strutted forks, long tank and rim brakes. A year later, in 1909, the Big 4 was joined by a smaller $3\frac{1}{2}$ hp version with fairly near square dimensions of 82 × 90 mm, giving a capacity of 475 cc. Norton rode a single in the 1909 TT but retired as he was to do in the subsequent two years, and in 1911 his design of the smaller machine was amended to the famous 79 × 100 mm dimensions and 490 cc. These figures were to ring down the years and bring instant recognition from owners and riders all over the world.

The two machines proved sufficiently popular to keep the company busy, so that plans in 1913 for a $2\frac{1}{2}$ hp model using a 70 × 90 mm engine of 346 cc capacity were shelved as had been those in 1911 for a more radical model to be called the Nortonette. This was to have used a two-stroke engine of 55 × 65 mm dimensions and 154 cc capacity and to have weighed but 60 lb.

In fact, the small company was running into financial problems due in part to Norton's idealism and lack of interest in money or profits. This resulted in liquidation in 1913 and a takeover by R. T. Shelley who were toolmakers and one of the principal creditors. Thus the company became Norton Motors Limited and this new arrangement allowed Norton to concentrate on his machines while the company became known and respected.

Up to 1911 few Nortons were produced and, aside from Fowler's TT victory, they made little or no impact on the sporting scene. In that year, however, they did make a mark, for Dan Bradbury was timed at over 70 mph in a speed trial on his Norton and this was the first time that a machine under 500 cc had managed this officially. Then in September 1912 Jack Emerson rode his Norton from Hull to Byfleet and ran in the 150-mile Brooklands TT where he came up against many very experienced riders. From the flag he was 2nd and soon went into a lead he stretched out for the rest of the race, showing the established Brooklands riders the way round. He finished over 13 minutes in front of the 2nd man, took three long distance world records in the process, and only missed the 3-hour record when he had tyre trouble.

After the event Emerson's engine was stripped and found to be quite standard apart from extra strong valve springs. It had been geared to run at just over 3000 rpm for the race. In case the Brooklands habitués thought it was a fluke win Emerson also rode in the Senior Hour ACU Championship at the October meeting and again finished well ahead of the field.

Emerson rode in the 1912 Senior TT but, like the other Norton runner, retired although the machines were reported to be extremely fast. This point was proved in July when Percy Brewster made his debut at Brooklands and set a new 500 cc class record for the flying mile at 73.57 mph. His engine was fairly non-standard, having been used in sprints and hillclimbs and internally featured a roller big-end, alloy rod and a light steel piston which with a special carburettor enabled it to run over 4000 rpm, a very high engine speed in those days.

Emerson nearly did it again at Brooklands in November being well ahead in the Senior Hour race when his float chamber top came adrift and he dropped back to finish 3rd.

So the marque became better known and James Norton looked ahead to alternatives for the four-stroke single and well beyond to alternative prime movers that would be far more efficient in their use of fuel. Steam turbines and radium energy were sources that crossed his fertile mind, while for the immediate future he conceived an overhead valve engine in 1913. In an interview he talked of 90 mph from the 500 cc side valve, 120 from the ohv, and 150 using desmodromic ohc. In time this was to happen, within a few mph, and yet the production models

Ralph Cawthorne at the start of the 1922 TT with the first ohv Norton which retired with many problems

Restored 1923 16H with classic lines and flat tank, fitted with later Amal

of this visionary man were never innovative in the accepted sense and were to hang on to the belt drive and the vintage frame longer than most.

The slogan 'Unapproachable' came into general use around 1911 but was first employed by Norton in 1908 to advertise a hillclimb success of Rem Fowler. It was to stay around for a long while and for many years was a justifiable truth rather than a sales slogan.

The takeover of 1913 brought a subsidiary benefit to Norton in the shape of R. T. Shelley's brother-in-law, ex-Singer works rider D. R. O'Donovan. Within a very short space of time his tuning ability enabled him to sort the Norton to his liking after which it was fast. He went from success to success to become a legend in the twenties as a tuning wizard, just as Joe Craig did in later years.

His activities led to the introduction in 1913 of the Brooklands model which was supplied stripped for racing with a certificate that 65 mph had been reached. From this was to come the legendary BS Brooklands Special and BRS Brooklands Road Special models of the twenties that represented the pinnacle of the side valve single with belt drive and flat tank.

In 1914 the engine was cleaned up by the use of new castings and it took the shape it was to maintain in outline for four decades. The frame up to that period had a straight top tube similar to a bicycle, but for 1915 the rear end was bent down to lower the riding position. O'Donovan kept to the earlier frame for this own records. One was the world kilo which was taken at 82·85 mph in 1915, and by that year the trademark on the tank had changed to the curly 'N' form it was to retain for all time, a great improvement over the earlier crude script.

Although by then the war was well under way production continued and in 1915 a three-speed Sturmey Archer gearbox became available with either chain or belt final drive. The gearbox did much of its testing fitted into a Big 4 as J. Cohen, the designer, was a Norton enthusiast and preferred the punch of a big single. This gave the box a harder time than the smoother power flow

Above **The Brooklands Road Special in 1922, just the very basics for motorcycling**

Right **Some of the 1927 range with Druid forks still in use and some very old-fashioned lines**

Model 14

Model 16H

Model 34

Model 19

odel 44

odel 1

odel 18

odel 25

of a vee-twin so was a better means of proving the product.

At the same time the Norton numbering system came into use in place of the earlier practice of just describing the machines by specification. Some confusion arises with these numbers, which run from 1 to 25 (less 23), but then changed to a letter sequence for four models before reverting to numbers with less logic and more gaps in the order 30, 34, 40, 44, 50 and 55. An early confusion arose because the Big 4 was called model no. 1, but at least the number was then retained for all time for a basic machine form. Thus no. 2 had the 490 cc engine and three-speed Sturmey box but the chain drive version was no. 16 and became in 1921 the famous 16H which continued as that until 1954.

1916 saw the factory move to Phillips Street which backed onto the Shelley works in Aston Brook Street. Later the Bracebridge Street site was added to complete a compact working unit and become one of the best known addresses in motorcyling. In the same year a young man named Gilbert Smith joined the company to devote his working life to Norton. He became a director in the twenties and managing director in 1945.

After the war the Norton range comprised five models little changed from 1915. The Big 4 was the largest but fitted with all chain drive, as was the model 16. For riders who preferred the silence and peace of the belt with just the click of the fastener they listed the model 9 using the same engine. For faster men there was the model 8 BRS guaranteed for 70 mph and fully road equipped within the meaning of the times, and for racing the stripped model 7 BS without brakes or silencer but good for 75 mph. This last came with the Binks 'rat trap' carburettor favoured for track work as it could not be eased off over the bumps. It was simply a long trumpet with completely clear tract and a jet. It could be closed off to give a choke effect to start, after which it would eight-stroke until it cleared and

ran full bore. To stop it the rider choked it to death.

The speed guarantee given with the BS and BRS models did not in fact apply to the machines themselves but to the engines. These were fitted to a slave frame O'Donovan kept at Brooklands, in early days it was *Old Miracle*, run in, dismantled and checked over, then reassembled and taken out for a timed lap of flying kilometre. Some seven a week would pass through this system and then go back to the factory to be assembled to their correct cycle parts. The

Bert Denly at Montlhèry on the occasion he put 100.57 miles into the hour. Nigel Spring (in hat) in centre

numbers were one reason that belt drive was favoured as it made the job much quicker, although O'Donovan had a good idea by then what internal shapes gave a quick engine.

The variation, which could amount to 10 mph, arose from the casting techniques of the times allied to the fixed cylinder head. If the combustion chamber roof sagged during the casting process the compression ratio was raised. Sometimes this advantage would be cancelled by the resulting poor chamber shape, but every now and then a high ratio and a good shape would come together. Such freaks would often become record breakers, although the reasons of flame front control, mixture turbulence and squish bands were still locked in the future research work Ricardo was to do. For these reasons the engines were never certified for a maximum speed, only that they had exceeded the required norm by an amount not specified.

In 1920 the marque returned to the Isle of Man keen to do well to augment their Brooklands reputation but still keeping to their policy of using standard models. These may have incorporated ideas for future production but then, and until Norton's death, they did not use special models or, at first, star riders. Later on the policy was changed and special racing machines appeared ridden by professionals. However, the notion of improving the breed was maintained and for years the works features would apear on the production model the following year and spin off to the tourers the year after.

The Senior TT of 1920 had a good Norton entry and Manxman Duggie Brown nearly pulled it off for them, taking the lead on the fourth lap but then dropping back to finish 2nd. Nortons also occupied positions 4, 7, 8, 10, 11, 13 and 14, so overall had a good day. The 11th man, Norman Black, had ridden a belt drive machine, while Graham Walker had a very tough and troubled ride to finish 13th in his first TT, the lowest finish he was to record over the years.

The TT engines had a single row crowded roller big-end which became standard the next year and ran with cast iron flywheels and pistons. The latter broke across the waist if the engine was over-revved while the wheels tended to work loose on the mainshafts. The works cure was to dip the shafts in sal ammoniac which gave a rust chemical bond for ever.

That same year Victor, or more correctly Vincent, Horsman came down to Brooklands with an all chain Norton he had prepared himself and walked off with the One Hour race at 71·68 mph to set a new world record. He then returned home to Liverpool and won a cup in a reliability trial held in North Wales using the same machine.

1921 saw Nortons back at the TT with shorter frames and revised gearboxes with bottom between second and top. This allowed the lever to be mounted low and booted positively down into second even if it did go through first on the way. The Sturmey was of a forgiving nature. Drum front brakes were also tried but so small as to be of no use and, while Dixon, Davies, Bennett and Edmond battled for honours on their new expensive special racers, the near standard Norton plodded round to finish 6th.

At the track O'Donovan, Horsman and Rex Judd went for the double 12-hour record and took it at 60·7 mph, while Judd became the first rider to officially exceed 90 mph on a 500 cc machine, with a speed of 92·44 mph over the kilometre. For this he rode the old O'Donovan side valve which had chain drive and a single gear with the gearbox shell carrying a countershaft only. With Rat Trap carburettor and high gear it had to be pushed down the test hill to get it going.

The 16H made its appearance that year and the derivation of number arose from a frame change. The model 16 used the Big 4 frame with high ground clearance, this version being known as the Colonial, and for 1921 appeared with a low or home frame based on that used in the 1920 TT. The Colonial model became the model 17C and the 16 took the initial letter 'H' from 'home'.

The side valve engine was fitted with adjustable tappets in 1922 so the need to file the valve stem to get the correct gap went, but the real Norton news that year was the appearance of an overhead valve engine. This continued to use the traditional Norton dimensions of 79 × 100 mm and was in effect a new top half on a very well developed bottom half.

It showed the artist in James Norton with simple classic lines and the gases going straight in and straight out of the cylinder head. The engineering was soundly based too with a well finned cylinder that was symmetrical apart from the recesses for the pushrods. The head was massive, well finned and carried its rockers well clear on pillars to give plenty of room for cooling air to flow over the combustion chamber. This made for a tall engine so that tank rail had to be kinked but, true to their ideals, Norton always kept their engines upright even when fashion dictated low lines, and other firms tilted their motors forward to suit.

The prototype of the new engine was sent to Brooklands for O'Donovan to develop and while at first he was uninterested in it, when finally fitted to a frame it raised the world 500 cc kilometre record to 89·22 mph. For the TT the firm played safe with only one of the new ohv models entered and this retired when the rider fell. Graham Walker was their best with a side valve at 5th.

The road model with ohv engine made its debut at the show that year as the model 18, a number that was to become as well known as the 16H and Big 4. With it was the belt-driven model 9 making its last appearance, but a rumoured miniature ohv single with 63 × 80 mm dimensions and 249 cc capacity failed to make its debut.

1923 was a good year for Norton and it seemed that the tardy final acceptance of valves upstairs had given the firm the breakthrough it needed. However, while the engine had power it also had some initial problems. One related to the valve springs, but deeply dished collars and longer springs overcame that, while another was head cracking cured by a dummy plug hole

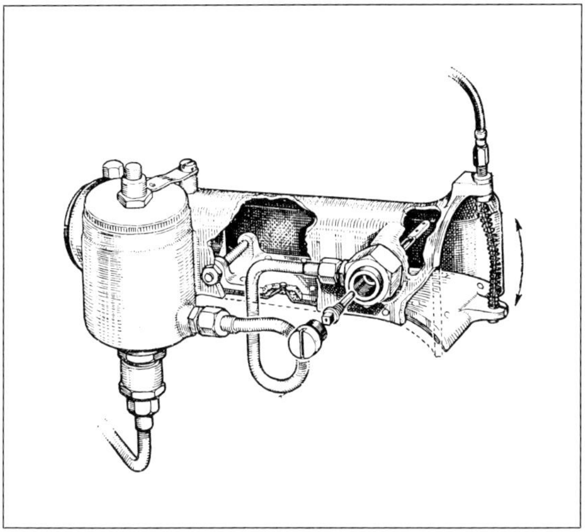

Left **The Binks Rat Trap carburettor, just a tube and a jet**

Bottom **Forecast of things coming, the 348 cc JE with rear magneto and external oil pump in 1931**

Below **The speedway Norton of 1930 with tuned ohv engine, Webb forks and empty gearbox shell carrying the countershaft**

between the pushrods to balance the working one. The cycle parts however lagged behind with a dummy belt rim at the back and a minute front drum. The side spring Druid forks did nothing for the handling and Graham Walker, although then Norton competition manager, defied threats of the sack and fitted Webbs. His demonstrations of their superiority during practice soon led to a wholesale switch by the other Norton runners.

A TT win still eluded them but Walker and Tucker were 2nd and 3rd in the sidecar race using extra long stroke 79 × 120 mm engines of 588 cc capacity, while Graeme Black finished 2nd in the very wet Senior with Walker 4th and Tommy Simister 5th. They all finished without brakes but the Sturmey box enabled them to slow down and they gained the team prize, an award which to Pa Norton was perhaps even better than a win as it demonstrated the consistent qualities of his machines.

These were further shown when Norton won the Maudes Trophy that year, the first time it was awarded. It was then called the Pettyt Cup after George Pettyt of Maudes Motor Mart, Exeter, who presented the ACU with a large silver vase to be awarded each year to the firm whose observed and certified test of a standard machine best demonstrated the qualities of reliability and economy rather than speed.

For their first attempt a model 18 engine was built from parts selected from the stores by the ACU at random and then installed in O'Donovan's track bike. His new rider, Bert Denly, did some running in and then the Norton ran for 12 hours to take that record and 17 others. It was a quick test to carry out in contrast to the drawn out road marathons to come in later years.

Brooklands also saw O'Donovan and Denly busy, with the hour falling to the latter at 82·67 mph in June and 85·22 mph late in the season. They and Nigel Spring captured a good few other records that year as well and just to confirm that the Norton was not only a speedster a team of three won golds in the ACU Stock Machine trial and another trio took golds home from the ACU Six Days Trial. Over in Ireland a local lad, Joe Craig, won the Ulster, the unlimited class of which he was to win four times in a row, and there were also successes in Europe.

If 1923 had been good, 1924 was magnificent and became known as the year of firsts. The first step the firm took was to appoint Walter Moore to design and build a machine to win the Senior TT. Moore came on his own terms, gave the 490 cc engine a good oil pump and fitted an adapted Ford drum brake to the rear wheel. The second step was to persuade Alec Bennett to ride for them. It worked and Alec won with an average which exceeded 60 mph for the first time while George Tucker, with Walter Moore as passenger to keep an eye on things, took the sidecar event on a 588 cc ohv single by half an hour. For the rest of that year event after event fell to the Norton along with the hour record to Denly once more at 87·07 mph, while in far off Italy a rider called Nuvolari won races on his Norton before moving on to Bianchi and then four wheels.

Once again Norton won the Maudes Trophy with a drawn out test using a Big 4 and sidecar. After the usual selection from stock and assembly it did the Lands End to John O'Groats run four times and then headed for Porlock Hill and a score of ascents. On route a charabanc came out of a turning and pushed the outfit up a bank into a wall, so overnight it was rebuilt using parts from another Big 4 in a dealer's showroom providentially close to hand. With new frame, forks, front wheel and chassis it completed the hill climbs and Norton retained the trophy.

That same year the standard models continued with their Druid forks but the works use of Webbs encouraged a change over that year or the next. Internally the big-end bearing became a double row roller.

The Norton triumphs were celebrated in Birmingham with a civic reception and it was fitting that kind, honest James Norton should be so honoured for he was a sick man. He died in April 1925 and it is a tribute to his faith and determination to put the best into the machines that bore his name that they continued to look like real Nortons nearly to the end of the single line.

1925 was not such a successful year. The machines had detail improvements carried out to enhance reliability but other makers were stimulated to more adventurous technical advancements and in the TT Norton's best was a 2nd in the sidecar and 3rd in the Senior. One innovation that year was a four-speed gearbox made by CAV with cross over drive so the output

Geoff Duke on *Old Miracle* at a Beaulieu occasion with Lord Montagu standing beside him. Lord Brabazon sits in the 1908 Austin racer

OK 290

sprocket was on the right which was useful with a sidecar. The box had hand change, the gear selection being done by a circular camplate, a type later adopted by Sturmey Archer. Later in the year it became an option for the Big 4 with a kickstart lever hung on the outside.

The Maudes Trophy was just asking for a third attempt so Norton obliged by running a 490 cc ohv solo and 588 cc sidecar through mock MCC trials plus the End to End run to tot up some 3000 miles. They then had a decoke, went to Brooklands and broke records up to 12 hours duration. They kept the trophy.

And held on to it in 1926 with a 588 cc ohv outfit which did 100 ascents of Bwlch-y-Groes in North Wales before the now traditional End to End run. Which again proved the Norton reliability but this picture was to start to crack up in the racing field that year. A new recruit joined the team, the great Stanley Woods, and with Bennett, Craig and Jim Shaw they were as strong as any of the opposition.

At first all seemed to go well, for Bennett won the French, and in the Island Woods came home in front with Craig 4th. However, the post-race examination showed that Woods' engine had had a hard time and would soon have slowed or stopped. In Europe the Norton men began to find that they would make the running to start with but then wilt and fall by the wayside.

It was all too apparent that the ohv engine had reached its limit or passed it, while the long, veteran frame that housed it was now 'too slow for occupant'. Something new was needed and in 1927 came both a new engine and a cradle frame that also housed the older ohv unit.

Norton were poised to take another of their few steps forward, as always after some deliberation. In the meantime the road range continued in its old format with a multiplicity of models generated by combining two side valve engines, two ohvs, two gearboxes and a miscellany of frames, wheels and cycle parts in just about any combination possible. Even Druid forks were still to be found on several models.

In 1927 Norton made their fifth and final attempt on the Maudes Trophy and again set out on a lengthy record breaking and road mileage marathon. Once again a motor coach capsized the outfit but the test continued with the machine running solo. Rather sadly the gudgeon pin broke in a final session at Brooklands and the trophy went to Ariel. Norton decided that the risky, complicated tests had gone on long enough and retired on the laurels of their four wins in a row.

The old pushrod motor still had a kick left in it and was of course to continue in use for many years. On the track it had perhaps its finest hour when Bert Denly went to Montlhèry and raised the 60-minute record over the ton for the first time using the 490 cc engine. It took four attempts before he managed the full period with magnetos dying in the closing minutes on the first two runs. Finally he managed 100·57 miles in the time and one of the toughest of all records went to an ohv unit for the last time.

The large singles were joined by smaller 350 cc units in 1928 and the super sports ES2 model had its magneto mounted behind the cylinder, chain driven from the left end of the crankshaft. Slowly the component parts of the later Norton single were moving into position but a year or two were to elapse before the final form was reached.

1929 brought saddle tanks but the road machines were now very dated in appearance and hiding the forward mounted magneto under a cast alloy shield or introducing twin port models in 1930 could not disguise the early vintage looks. As late as 1929 the 16H still had Druid forks plus a curious twin barrelled silencer that did little for its appearance.

It was a period of marking time and during it, in 1930, a speedway model was produced using modified stock parts. It looked short and businesslike but failed to catch on.

Behind the scenes a complete redesign was going on.

2 'Camshaft'

The new works engine for 1927 was the overhead camshaft type designed by Walter Moore. Following on the success of the revised 1924 models he learnt that other firms were thinking of moving the camshaft upstairs so set out to design such an engine in his own time. The results of his labours were shown to Norton and in due course they agreed to try the new idea.

Two engines were built up but not completed until TT practice was underway and the works originally intended to keep to the tried and tested ohv motors. They had these in the new cradle frame which set a pattern that continued with minor changes for two decades or more and truly did improve the steering and handling to make the machines 'unapproachable' through the bends.

The camshaft Norton retained the traditional 79 × 100 mm engine dimensions of the marque and the timing side of the engine was altered to provide a housing for the bottom bevel pair of the vertical camshaft drive. Below the gears went a Moore-designed reciprocating oil pump. This had first been used in an adjustable form in 1925 but on the ohc engine the output was fixed but an external control varied the pick-up level of the scavenge side. When set on 'High' the oil level in the crankcase was allowed to rise enough for a semi-wet sump effect to occur.

The bottom bevel of the vertical shaft ran in a ballrace mounted in a housing bolted to the timing side. The shaft was splined into it with the top bevel keyed in place to drive the camshaft. This ran in a separate cam box carried on the cylinder head in the same manner as the ohv motor with the rocker shafts extending out of the box with the valve actuating arms attached to them on keyed tapers.

The cam box had an inspection cover over the top bevels and the drive shaft enclosed by a tube with large gland nut attaching it to the underside of the box. Its lower end went into the timing cover which was characterized by a blister cast to give clearance to the mechanism. An oil pipe ran from the front of the timing chest to the cam box in which the two cams were separately mounted on the shaft.

The remainder of the engine followed the lines of the ohv unit with suitable changes to accommodate the camshaft drive. The cylinder head thus had only one cut-away at the side to provide clearance, while exposed valves with coil springs continued. Ignition was by magneto mounted on a platform behind the engine and driven from the crankshaft by a chain running on a sprocket mounted inboard of the main drive one on the left. The exhaust pipe went to the same side of the machine and ran back straight for most of its length. The gearbox had three speeds and the control lever laid right down as before so that it could be worked by the foot despite still being hit and miss for the middle gear. The final drive chain was on the left.

The frame was built up from tubes and lugs with the down and seat tubes running into a cradle beneath the engine. Triple rear stays supported the rear wheel and made sure it remained in line, with the front one carried in Webb girders. It was typical Norton, big, massive and strong. It also steered.

The wheels carried large drum brakes and the frame a true saddle tank, unlike the 1926 one which looked like a saddle but was in reality two pannier tanks clamped on to the frame and together. The oil tank was mounted on the seat tube and behind it on the centre chain stay went a small toolbox. The works machines had a lifting handle formed in the rear mudguard stay and a rear stand, the first to enable the second to be used.

At that time the camshaft engine was undeveloped and many such designs were slower than the ohv units they were copied from. It was to be a while before work with a stroboscope, carried out by Velocette, showed how little the ohv layouts followed their actual cam contours and how well they kept to them with ohc. With that knowledge came cams with overlap and the end of the racing ohv.

The 1927 TT debut of the camshaft engine saw Stanley Woods go straight into the lead he held for five laps. When he re-fuelled no one told him his lead was four minutes so he continued unabated and his clutch went. Alec Bennett then moved up to finish eight minutes ahead of the next man and his clutch failed when he tried to restart after the race. Stronger internal tongues fixed that and Woods went off into Europe and a successful tour of the major events.

It was no great surprise that at the show Norton had a new model for 1928, the CS1, or Camshaft One, based on the works machines. Alongside it was the ES2 sharing the same frame and cycle parts. The camshaft engine continued the tradition of lofty motors in tall machines and still retained the vintage look with just a hint of the changes to come.

1928 brought a pair of smaller Norton engines both using 71 × 88 mm dimensions to give a capacity of 348 cc. One had ohv but the other was a replica of the CS1. Differences were minimal with the ohv motor having a rocker box similar to the camshaft type and both engines using a lightened version of the cradle frame without the middle chainstay and fitted with lighter cycle parts.

It was not a good year for the firm at the TT and the new Junior models had a disastrous day, all five retiring. In the Senior things were little better, although Joe Craig was at least in the lead when

Above **1928 CS1 with Moore engine in cradle frame**

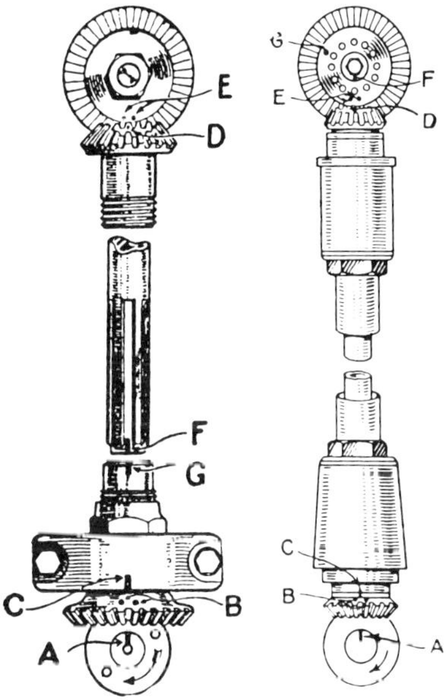

Right **The Moore (left) and Carroll camshaft drives with timing marks**

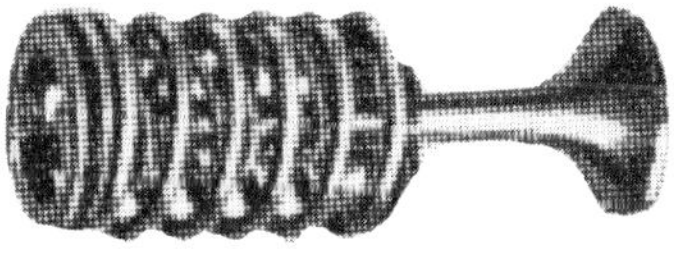

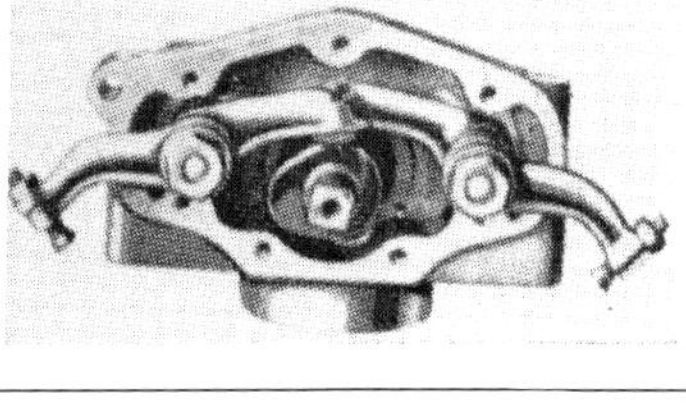

Below **The Moore camshaft engine and its head, valve and cambox less side cover**

Above **Carroll Norton, at a race meeting after the war**

Top **Moore engined machine also at postwar race meeting, fitted with footchange mechanism**

Right **Over restored Moore CS1 at Ramsey during 1975 TT weeks. They never shone like this for real**

he fell and retired. In Europe things went better and some successes came to the make.

Encouraged, the firm made few changes to the works machines for 1929 and the most obvious was that the exhaust pipe was moved on to the right side to leave the clutch and transmission more readily accessible. The more important change was the adoption of a positive stop mechanism for the gearchange following Velocette's lead the year before.

It was again not a good year for the company with no Junior TT finishers and only a 4th in the Senior, while in Europe success was limited. As had been expected, the two 350s were introduced into the range, the pushrod model being typed the JE or Junior with Enclosed pushrods and the other the CJ or Camshaft Junior.

That year Walter Moore left the company and joined the NSU firm which had been after his services for some years. Initially he was not interested in their overtures but when his

expectations of a seat on the Norton board appeared to wane he was approached again and asked to name his own price. To his surprise it was accepted and, when he arrived, he discovered a large house with garden and gardeners went with the job. As the camshaft engine was his own design he was able to use it again in Germany which resulted in the NSU looking just like the 1928 works Norton, but with a four-speed gearbox. It had considerable success in its own country.

In place of Moore came Arthur Carroll who, assisted by Edgar Franks, set about a complete redesign of both road and racing engines. Carroll loved motorcycles and riding them fast which was to bring his death in 1935 in an accident, but not before he had established the format of all Norton singles to the end of the line. Franks was more concerned with the draughting of the road models and in later years became well known to generations of Norton owners for his clearcut instruction books. With them worked Joe Craig as development engineer and race chief, a pair of roles he was to become world famous in.

The new overhead camshaft engine became dubbed the Carroll engine and owed little to the earlier Moore unit. Carroll did what most in his position would, had a good hard look at the opposition and used the features that appealed. It is also very likely that his brief included an instruction to drop as many of the Moore features as possible and to make the appearance of the new engine completely different.

So out went the reciprocating oil pump, the splined vertical shaft, the oiltight cambox and crankshaft magneto drive. In their place came the most famous ohc motorcycle engine of all time, the camshaft Norton. In fact there was a good deal of similarity between the Carroll engine and the Velocette with the vertical drive shaft, oil pump location and magneto drive all rather KTT However, it had its own distinct Norton air of massive ruggedness whilst looking completely different from the Moore engine.

Perhaps the truth is that if the camshaft is driven by a vertical shaft rather than by chain or gears then there are few options as to the position of parts and the layout of their drives.

The Carroll engines retained the same dimensions as their predecessors to achieve their 348 and 490 cc capacities and followed the tried and trusted Norton philosophy of simple, tough construction.

The crankcase was split vertically on the cylinder centre line and the crankshaft turned in

two bearings on the drive side and one on the timing. The crankshaft was built up with a double row caged roller big-end with the crankpin pulled up against shoulders by good sized nuts. The rod had a bushed small-end and carried a conventional piston on a circlip retained, fully floating gudgeon pin.

The piston ran in an iron barrel which spigoted down into the crankcase and up into the iron head. Both these were retained by four through bolts which attached to long studs in the crankcase. Into the head bolts screwed four more bolts to secure the cam box so that this assembly was easy to remove for attention. The valves were closed by coil springs.

The cam box was split vertically with the single camshaft running in a ball race in each case half. The driving bevel had a pin and vernier hole arrangement locating it to the shaft and the exhaust cam was keyed to the shaft, with the inlet pinned to the exhaust by a roller. Thanks to the bevel vernier it was possible to alter cam timing very quickly and easily.

Tim Hunt (standing) and Jimmy Guthrie with the Junior model in 1931 when they finished 1–2 in both TT races

In time this was to give some tuners a useful bonus in their work as, when the camshaft Nortons were sold in the 1930s, they left the works with a cam timing suitable for an open pipe. On the road or with a Brooklands silencer the performance dropped but could be retrieved by altering the bevel vernier which was easy enough to do. A good few pounds were made that way on the basis of £1 per mph increase.

The cam box carried the two rockers which were straight with adjustors at their outer ends.

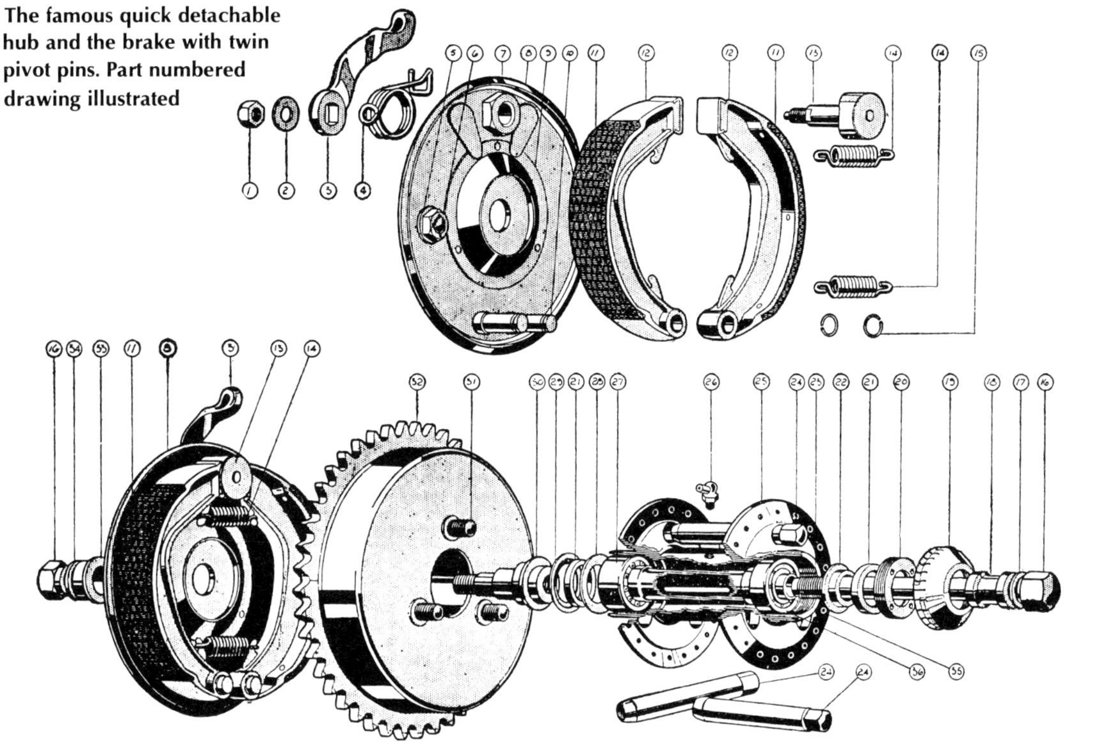

The famous quick detachable hub and the brake with twin pivot pins. Part numbered drawing illustrated

Like the Velocette, the outer ends protruded through slots in the cam box, and like that firm Norton never succeeded in making the arrangement oil tight despite a variety of seals and packings.

The camshaft was driven by a vertical shaft containing two Oldham couplings to deal with any misalignment and thermal expansion and an upper bevel gear carried in a housing bolted to the underside of the cam box. A similar housing bolted to the crankcase and carried the lower shaft bevel, while the drive shaft was enclosed in a tunnel attached to both housings by large gland nuts.

The bottom bevel meshed with another keyed to the crankshaft out-board of which and located by the same key was a spur pinion. This drove a gear and these three lived in a square housing cast into the side of the timing side crankcase, the top surface of this housing carrying the lower vertical drive bevel. Behind the second, and larger, gear went the oil pump which was driven by a plate that located into two holes in the gear and fitted onto the pump spindle.

The pump itself was of the gear type with three pairs for pressure, scavenge and bottom bevel box scavenge all in a circular housing held together by two screws and in the crankcase by two more. A pressure release valve was fitted into the rear side of the casting and a supply to the cam box taken from the top of the crankcase, this draining down the vertical drive shaft tube to the bottom bevel box. Small pipes took a little oil from the top bevel box to the valve guides, the exhaust one of which tended to become blocked with carbon. The two bevel boxes were joined by a pipe on the first models but this was deleted the following year.

The gear housing on the crankcase was closed by a lid that extended back to the magneto which was mounted on a platform cast onto the rear of the crankcase. The oil pump drive gear had its spindle extended through this lid to provide itself with a bearing and a small sprocket was fixed to the outer end. This drove the magneto by a chain which was enclosed by an outer cover fixed by a pair of screws.

This engine went into the existing cradle frame with three rear stays and drove a three-speed, footchange, Sturmey Archer gearbox without kickstarter. The plain exhaust pipe ran on the right past the end of the gearbox and tucked in against it below the external change mechanism.

The forks continued to be Webb with a single compression spring and adjustable friction dampers at the front end of the bottom links. A friction steering damper sat above the forks and both brakes were single leading-shoe drums mounted on the right side of the machine so the rear one was remote from the sprocket. A rear stand was still fitted along with the relevant lifting handle and rider comfort was provided by a saddle, a pad on the mudguard and kneegrips you could really hang on to.

The petrol tank was the saddle type and it continued to have the scolloped edge used to increase the joint area between base and sides. The oil tank was mounted on the seat tube with the filler on the right side. Mixture was supplied by an Amal carburettor which was the clip fitting type.

The Carroll and Moore engines were combined at one point to make a mongrel, possibly for show purposes. The compromise used the Carroll bottom half complete with a Moore cylinder head with the exhaust port and pipe on the left. None were believed to have been sold at the time but one such model was created for vintage racing in later years.

The smaller camshaft machine, the CJ in 1931. Cradle frame with two rear stays, rear brake and exhaust on right

The Carroll engine was very nearly right from the word go but 1930 was not to be its year. The Rudge machines were better and there were teething troubles. However, Jimmy Simpson failed to break his Senior model and scored his first finish in the 500 cc TT with a 3rd place.

Stanley Woods had some success in Europe and then appeared at the Ulster with a modified engine with downdraught inlet in a frame with only two rear stays. This was an interim machine that tried out some of the features intended for 1931 and they proved to be successful for Woods won. Thus there was not too much needed to be done over the winter when the final hone was put on the design and Norton began their domination of the 1930s racing scene.

The road models began 1930 with the Moore engine in the CS1 but by the Manx GP a few Carroll units were to be seen in the three stay frame. The CJ continued with the two stay frame that it always used. The smaller machine used the right side exhaust port and system while the 500 retained the left one and this arrangement continued for 1931. In 1932 all the camshaft models adopted ports and exhausts on the right where they were to stay.

New features on the 1931 works machines were engines as used by Woods in the 1930 Ulster but with no pipe connecting the two bevel boxes. The frame was much as in Ireland but altered to take the left side rear brake used from then on for many years. An André steering damper was fitted along with four-speed gearboxes, 14 mm plugs and slipper pistons. The machines had success after success in the major road races that year so there was no doubt that the development work had been on the right lines. The CS1 and CJ with Carroll engine continued with a shorter wheelbase frame with triple rear stay and lower saddle, two stay for the smaller model, three-speed gearbox and handchange as standard and positive stop footchange as an option. Pure racing machines were not listed but a number of replicas were supplied to known faster riders. At that time there were in fact few differences between a fully equipped CS1 with lights, silencer, rear carrier and toolbox, and a works machine. A racing magneto replaced the magdyno, some items were removed and the head and cams might have differed, while a close ratio gearbox without

Bottom **The 1935 model 40 International with the Norton gearbox and fitted with lights, the headlamp supported on two prongs**

Below **By 1934 the CJ had the check spring forks and the Norton primary chaincase**

kickstarter would be fitted.

The works Norton that had won the 1930 Ulster was tested by *The Motor Cycle* not as a road burner but over the West Country trials hills of the day. The report gave no indication at all of the machine's speed capabilities—a curious way of approaching such a machine—but all in line with the editorial policies of that era.

There was no gainsaying the success of the Norton, however, and after the successful racing season the point was rammed home by Bill Lacey who took the world hour record at 110·80 mph at

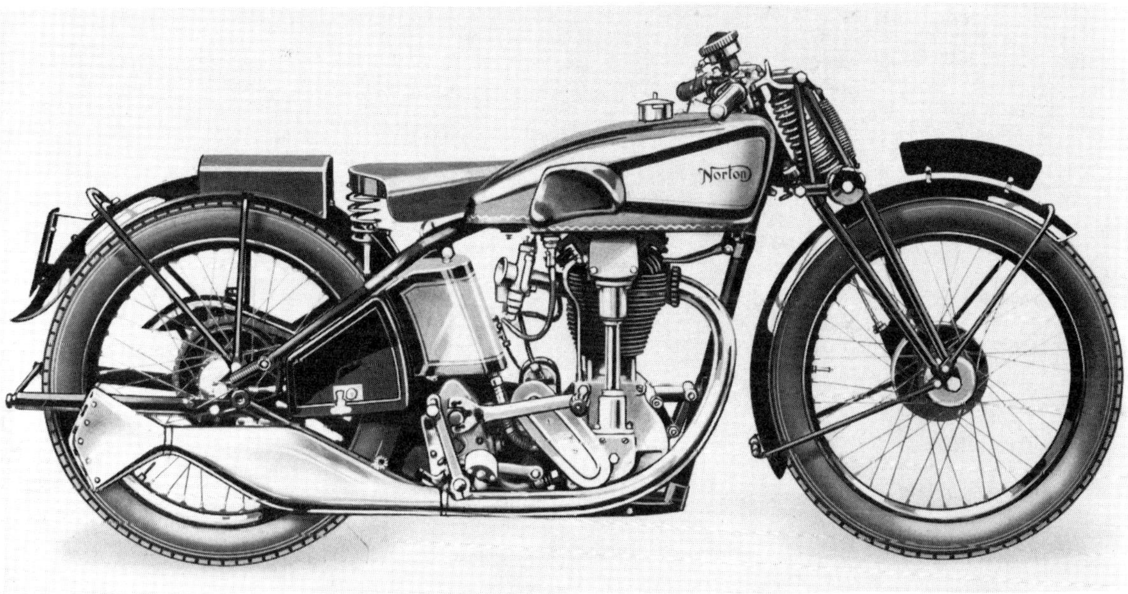

Montlhèry. Bert Denly, the holder, helped him by signalling his progress.

For 1932 Norton decided that their works machines should lose some weight by making many of the parts in alloy, while the engine finning was increased and the front forks were

Above **Model 40 International in 1933 less lights but with check springs and fantail silencer**

Top **1936 model CS1 with subtle variations to the International**

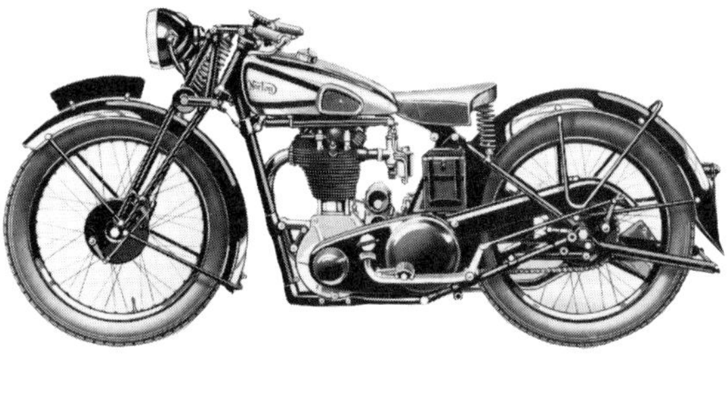

given check springs. These were connected between the forward extended fork links which were angled to give the desired effect. On the racing machines these springs were of parallel form and a fair diameter. The works machines had a narrow front hub that, unlike the road models, was not qd or interchangeable although it could be removed from the forks easily enough. The hub and brake drum were in one piece.

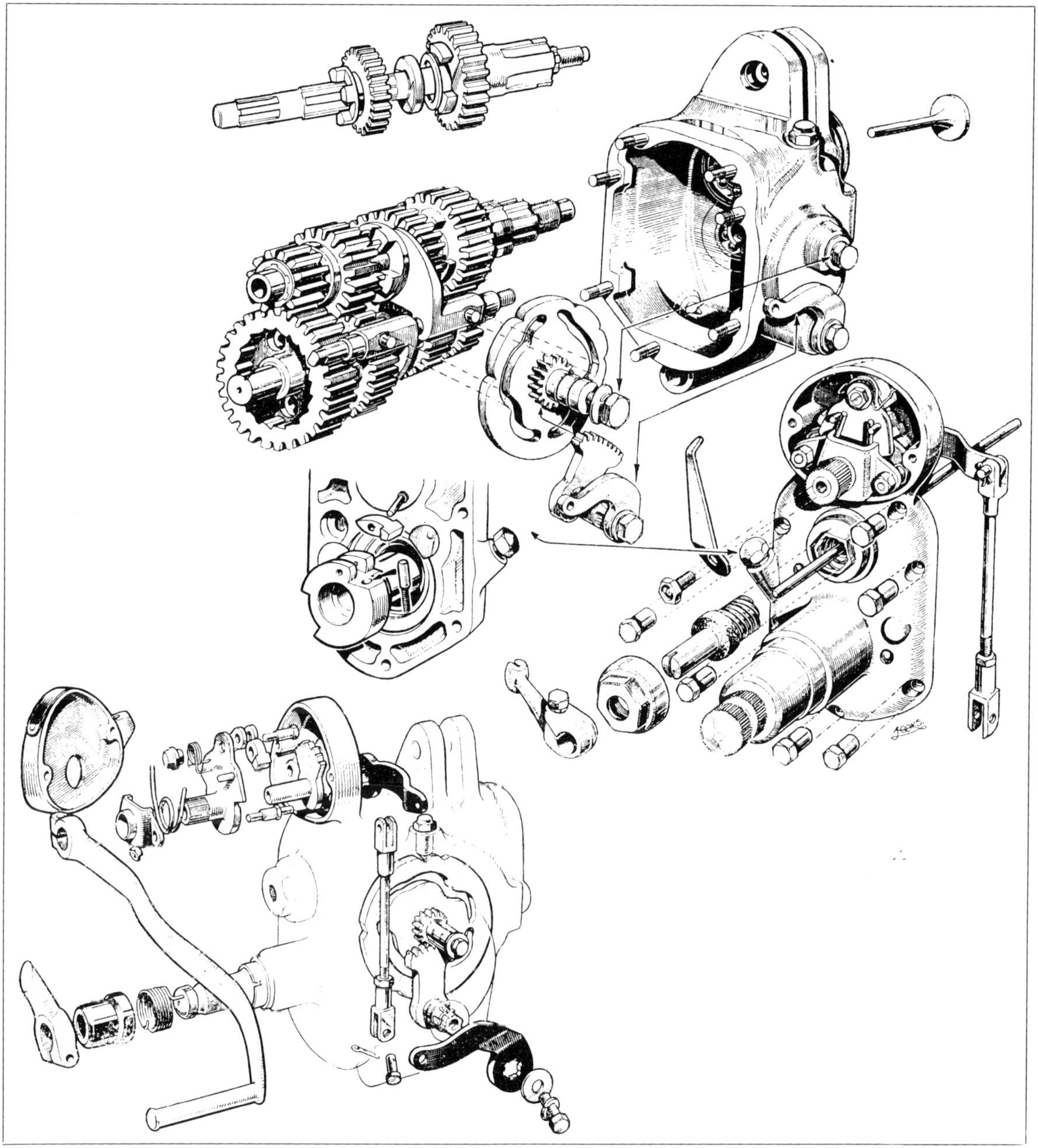

Above left **Jimmy Guthrie at Governor's Bridge in the 1934 Junior TT he won by 9 seconds**

Left **1933 model CJ with 1934 chaincase**

Above **The famous Norton gearbox. Postwar it gained an end cover and later the mechanism moved to give the laid-down type from which the AMC box was derived**

Above **Jimmy Guthrie before the 1935 Dutch TT, Joe Craig behind the pit counter**

Right **Start of the 1935 350 cc Dutch TT with, aptly, the NSU (33) ridden by Ted Mellors alongside Walter Rusk on the Norton (39)**

For the private owner the great news that year was the appearance of the International in both engine sizes alongside the CS1 and CJ which continued but more in a touring guise. The Inters were for racing and were replicas of the works machines less the developments of the current year and with some road machine features. Thus they had the qd wheels and, although fitted with racing cams and pipes, could be had with lights. There were alternative internals available for the gearbox which had four speeds and kickstart or not, depending on the ratios.

This point was determined by the design as the kickstart worked inside the bottom gear on the layshaft. With the close ratio box this gear became too small to accommodate the mechanism so any Norton with a close box had to be push started.

The wheels were the interchangeable Norton design with three fixing studs that were to be used for the next quarter century. Earlier work with another quickly detachable design had not proved successful as the wheel was prone to detach itself all too easily despite a locking device for the wheel nut and another for the locking device. The design that was used comprised a hub with equal size flanges held against the combined sprocket and brake drum by three long sleeve nuts working on conical rebates in the hub. Tight, it worked well but if run slack it soon wore and would then never stay put.

The CS1 continued with its three-speed gearbox but with the option of a four-speed, hand change, unit if desired. It also had the road-type mag-dyno in place of the BTH type KD1 racing magneto fitted to the Inter.

An Inter was tested by *Motor Cycling* early in 1932 but there were few hard facts in the report. A note that it could be ridden hands off at 90 gave an indication of performance and handling, but that was about it. Fortunately the machine was also taken for a timed test with Jimmy Simpson riding it and he put it through the quarter-mile at over 90 and, with the silencer removed, managed the level 100 in one direction. A later test by *The Motor Cycle* on the machine complete with its silencer and mag-dyno resulted in 85 mph which was still a fair speed for a road 500. In fact the Inter was always sensitive to the addition of a silencer and sold with racing cams suited to an open pipe. The absence of ramps also meant that the valve gear always clattered a good deal.

1933 saw the works machines fitted with aluminium bronze cylinder heads to improve the thermal conductivity, even if it did nothing for the weight. The cylinder continued in iron and the machines had another successful year.

The International models were fitted with Norton forks but altered from the earlier type although still much based on the Webb. A feature of these had been the knurled thrust washers on each spindle which took the side loads and helped when adjusting the end play for zero but free to turn. These reappeared and the fork links were extended forward for check springs of barrel type. The linkage was designed by Carroll to supplement the main spring when it was most wanted at extreme deflection and rebound. Minor changes affected the paint finish and the oil tank which was given a longer neck. The touring CS1 and CJ were given the four-speed gearbox as standard, although the positive stop mechanism remained an option. Their silencer was amended from fishtail to fantail.

The next year, 1934, saw the works team just as successful with further modified engines. The most noticeable alteration was the adoption of hairpin valve springs, while the head itself was changed to bi-metal construction with aluminium fins cast around a bronze skull. The cylinder fins also became aluminium with an iron liner and some machines had twin sparking plugs fired by a special twin-spark magneto. The second plug went in behind the cam drive and could only be changed if the cam box was removed. These engines had megaphone exhausts. On the cycle side the tank fixing was changed to the bolt-through type with four long vertical bolts running through rubber bushes set in tubes fitted into the tank to fully insulate it from the frame.

There were also changes to both the Inters and the CS1 and CJ. The oil supply to the cam box was increased and this in turn meant that the return holes in the vertical shaft housings had to be made bigger. A washer was added at the bottom of the shaft to catch any errant balls from the bearings and prevent them from reaching the oil pump, while a filter bolt went into the bottom bevel chamber.

The gearbox was given a much neater positive stop mechanism, the Norton forks went onto the C models, while the Inter was fitted with the one-piece hub and drum front wheel used on the racing machines together with the narrower front fork to suit it. The C models were fitted with tubular silencers.

The transmission received a major revision that year with the adoption of the Norton clutch and oil-bath primary chaincase. The clutch was the first part of a change from Sturmey Archer parts and was developed to reduce costs. It was a straightforward multi-plate design clamped up by three springs and had a shock absorber built into the centre hub. This was efficient enough to delete the need for one on the engine mainshaft so that went to further reduce costs. The clutch was lifted by a large diameter mushroom which was one of its secrets for the pressure plate had little chance to do anything but lift square. This ensured the clutch separated well and was half the key of the good Norton gearchange to come. The other half was the gearbox itself but that was still a year away. The clutch was lined with Ferodo inserts developed to run in oil but also had a band around the drum to keep the amount

Camshaft model from the middle thirties, a 1936 CS1 or CJ

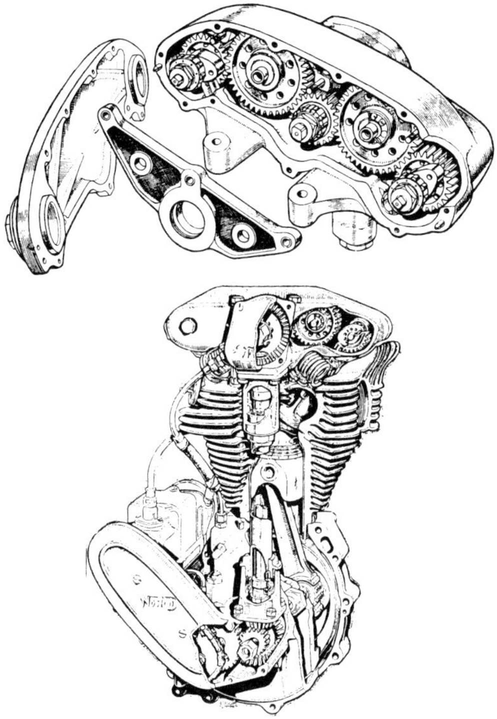

The double knocker engine and details of the cambox

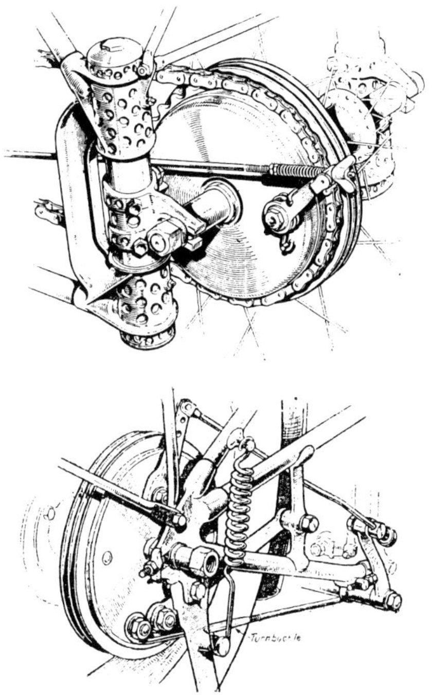

Works plunger as used in 1936 and earlier remote brake in rigid frame

down to an acceptable level.

The oil bath was like most good designs, simple in its concept. It comprised a pressing behind the transmission with a wide ledge running round near the edge, a wide rubber band placed round the ledge, and an outer cover that fitted over the band to seal against it. A single large nut on the footrest hanger kept it all in place and it was very tolerant of abuse such as being leant on a kerb to park the bike. An inspection cap enabled the chain tension to be checked or oil to be added and a single bolt gave a level.

The works Nortons continued their triumphant progress through 1935. The Inters were offered with hairpin valve springs as standard and aluminium-bronze cylinder heads as an option, although it was suggested that while these were fine for racing the valve seats were prone to sink so the cast iron head was to be preferred for road use.

All the camshaft models were fitted with the four-speed, footchange Norton gearbox that year which was based on the old Sturmey Archer unit and was to continue virtually without change for many years. The gearbox came into being because the Sturmey boxes were made by Raleigh and, when they left the motorcycle market in 1934, they stopped gearbox production. Norton had used their boxes from the time they changed from belt drive and were disinclined to use Albion or Burman units as did other makers, so arranged to take over the Sturmey design and have it made by Burman. Before manufacture began Norton modified the

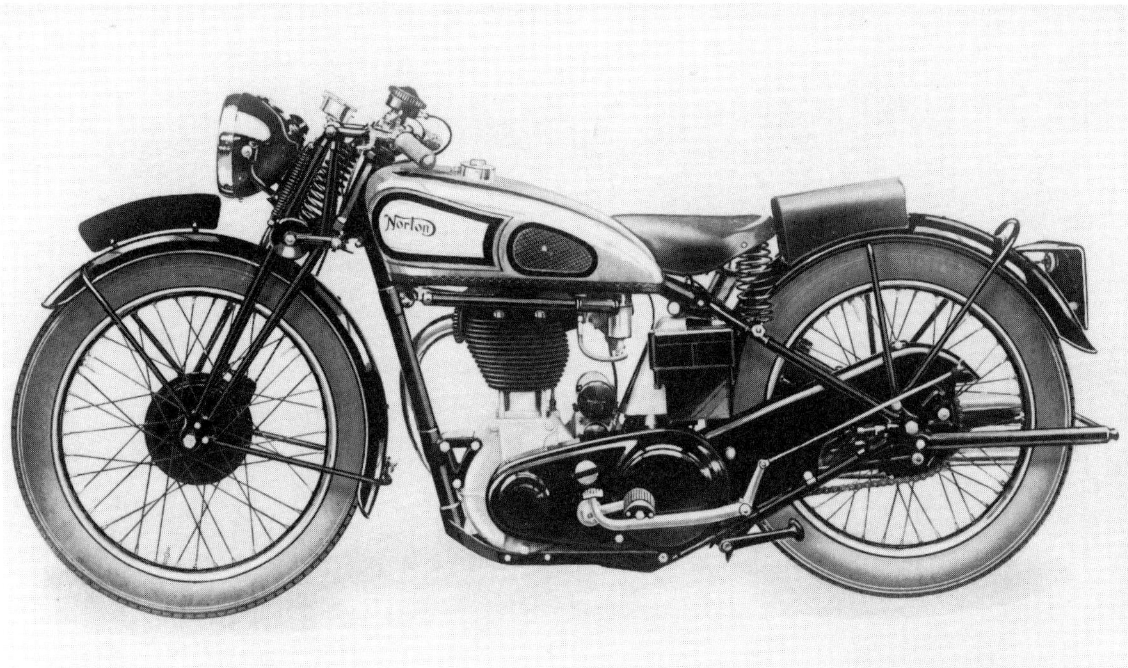

Above **The 1938 International model 30 with narrow mudguards, long tank and full equipment**

Below **1938 model CJ, rather less sporting than the Inter but with the same strong camshaft engine**

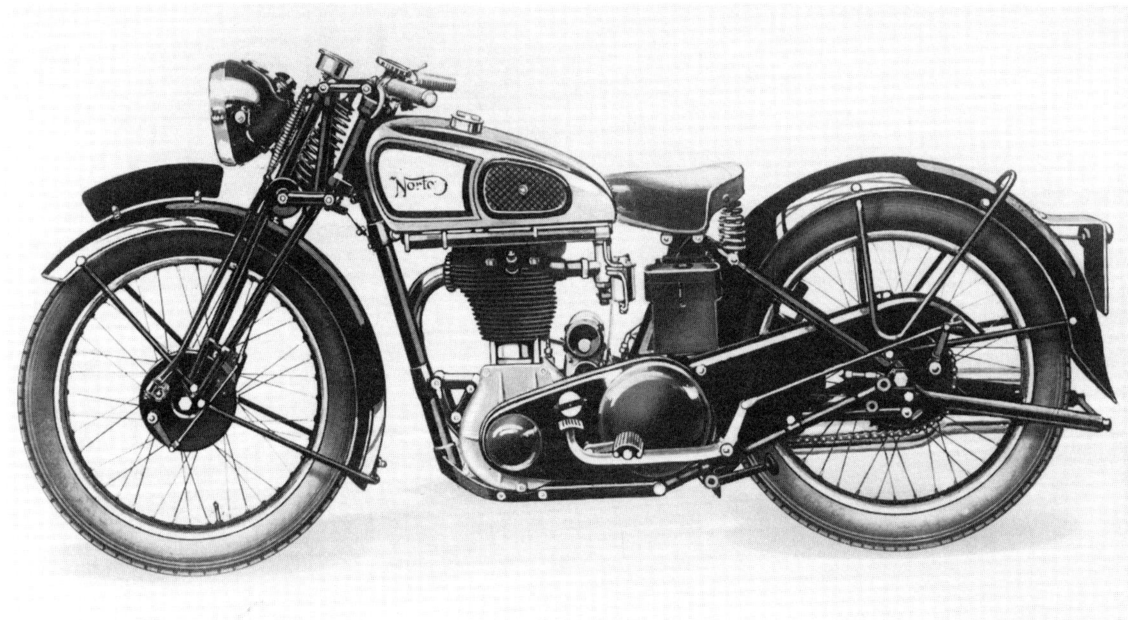

design to fit a row of rollers in the sleeve gear to help support the mainshaft, a roller bearing for the layshaft, and redesigned the positive stop mechanism.

The box was a straightforward design with mainshaft concentric with sleeve gear mounted above the layshaft. Clutch and final drive both went on the left, the first out-board of the second, and the clutch pushrod ran through the mainshaft. Each shaft carried four gears, one of which was splined to move sideways under the guidance of its selector to engage with the gears either side of it with three or four dogs and plenty of backlash. The layout of the box was such that at each gearchange one gear moved out of engagement and the other then moved in, a feature which along with the ample backlash for the dogs and the clean lifting clutch made for a superb gearchange seldom equalled over the years.

The two selector forks that moved the gears were carried on a single rod and moved by a circular camplate pivoted in the front wall of the gearbox and located by a spring-loaded detent plunger at the top. The plate was geared to a quadrant whose shaft ran through the front of the shell and carried a small external lever. This was linked to the positive stop mechanism mounted in an upward extension of the gearbox end cover. The link holes and pins were prone to wear which would throw the change into confusion, but the holes could be drilled and clevis pins from the front fork check springs used. Being larger and harder they would take the load more easily without wear.

The end cover carried the kickstarter spindle which supported the layshaft, the combined filler and level plug for the lubricant and a ballrace for the mainshaft. This was held in place by the clutch worm housing and the worm was turned by a small lever clamped to it and connected to the clutch cable whose adjustor went into a lug cast onto the small cover of the positive stop mechanism.

The main gearbox casting had substantial lugs top and bottom for mounting and the upper were prone to fracture. The box pivoted about the single lower mounting for chain adjustment and this lug went between the frame members without any problem. The top lug was divided into two and fitted either side of a frame member with a single clamping bolt. If left slack movement would wear the aluminium away so that re-tightening would crack one or both lugs off.

1936 was again a successful year for the works machines which were fitted with plunger rear suspension, the parts of which were well drilled to reduce weight. This system was first tried at Brooklands and had no damping. At the front the fork spring was mounted on spherical seats at both ends to remove the bending loads occurring on full deflection. The bore of the larger engine was taken out to 79·6 mm to increase the capacity to 498 cc and the dual ignition continued in use. An experimental Bowden carburettor with multiple jets was tried but found wanting.

The International models were given the option of the bi-metal, aluminium finned, bronze skull head and the light alloy barrel, while all the camshaft engines changed to a 14 mm plug that year.

1937 brought a further step forward for the works models with the appearance of twin overhead camshafts. Below the cam box the engine remained as it was, apart from some more finning on the head and barrel plus the abandonment of the twin plug ignition system. The new method of valve operation involved a train of five spur gears in an extended cam box with the centre one driven via a vernier coupling from the top bevel as before and the two outer ones carrying the cams. Again vernier couplings were used to connect the cams to their drive gears. Short tappets went between the cam and the valve stem to take the wiping forces of the cam on the bushes in the cam box. Hairpin valve springs continued in use and apart from these

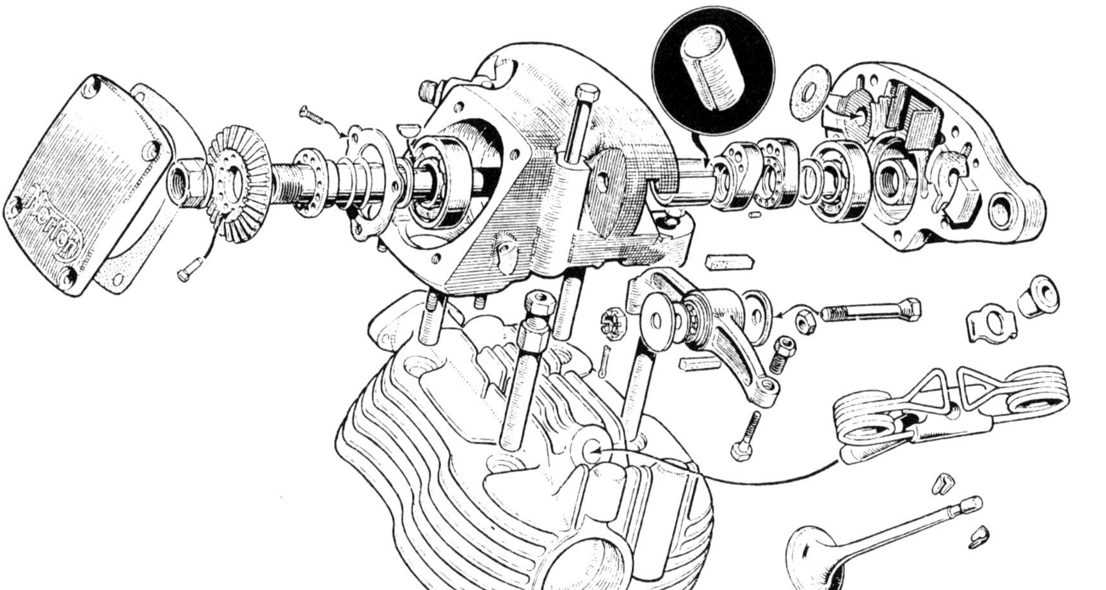

Construction of the single camshaft head and cambox, designed to leak oil

and the valve stems, the valve gear was totally enclosed.

One machine did appear in the TT with full valve enclosure achieved by using coil springs and a new cambox with separate rocker covers. It had been designed by Edgar Franks but was abandoned because broken springs could not be replaced easily and in seconds in a race as could be done with the exposed hairpins. On the cycle side, the works machines were fitted with light alloy rear sprockets and a rev-counter driven from the lower magneto sprocket by an extension that coupled to a right angle drive box on the outer cover.

For 1938 the works motors were revamped in many areas, although the overall concept remained unchanged. To enable the engine to run faster without an excessive piston speed, and to help the breathing with bigger valves, the bore and stroke was changed on both engines. The smaller became 75·9 × 77 mm to give 348 cc,

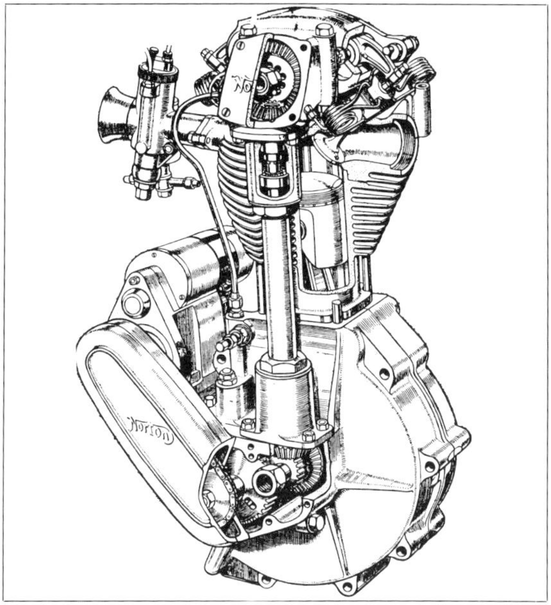

Above **The single camshaft road engine with valve lifter and mag-dyno**
Below **The Junior 1939 Manx Grand Prix, this one a superb restoration**

and the larger 82 × 94·3 mm and 498 cc.

The bigger valves went with a larger carburettor and, at the other end of the process, a larger megaphone. The cylinder head had bigger fins and these took on the square format, while the cylinder was sunk deeper into a new crankcase which sported vertical cooling fins for the first time. The lower bevel housing was altered to allow the exhaust pipe to run higher and the result was about 50 bhp at 6500 rpm running on 50/50 petrol/benzole, sufficient to propel the 500 at around 125 mph. Although a big single, it had a fairly narrow power band and would fall off the megaphone fairly easily. The engine, in both sizes, had to be warmed up very carefully and any attempt to use all the revs early in a race usually resulted in a seized piston.

The new engine was given a modified frame that had lost its forged cradle under the power unit. In its place the two lower chainstays ran from the bottom of the rear plunger units to the down tube.

The suspension was changed at the rear with a much smaller and neater arrangement of the plungers with forged light alloy sliders. Even more noticeable was the change at the front where telescopic forks were fitted. Although these looked much like the hydraulically damped BMW type they in fact had main and rebound springs only without any damping. The initial reaction of the riders was not too favourable but tests at Donington showed the new forks to produce shorter lap times, due it seemed to their improving the performance of the front brake.

The brakes went into conical hubs cast in a magnesium alloy and the drums were ribbed. The front brake had an air scoop attached to the backplate.

Such was the works Norton in 1938 and in the Senior TT Harold Daniell set the seal on its fame with his 91 mph record last lap which stood for 12 years. His race speed was also a record, but in Europe things proved to be more difficult, for the blown BMWs had become hard to catch even if

the more agile Norton could make up yards round the corners.

For the private owner the Inter was offered the option of the plunger frame and for the few who took this up there was a non-qd rear wheel to match. This retained the separate hub, sprocket and drum, also the brake, but had a one-piece spindle for the assembly to rotate on.

At the end of the racing season Norton announced that they would not be running a team in 1939 as they were too busy with orders for their road models which included substantial military contracts for the 16H. Early in the new year a further surprise came when Joe Craig left Norton and went first to BSA and then on to AJS.

1939 was the year when the supercharged engine came into its own with BMW winning the Senior TT and Gilera the Ulster, while blown machines were run or tried or built by AJS, Velocette, Guzzi, NSU, DKW, Benelli and Bianchi. The obsolete single still managed places in the TT but in Ulster Frith was well discouraged when absolutely flat out on his Norton and really carefully tucked in, to have Serafini bring the Gilera alongside, look at Frith's rev-counter, grin, wind on the rest of the throttle and disappear down the road. All done without getting down to it either!

For the private owner 1939 was notable for the appearance of the Manx Grand Prix model which, postwar, became the Manx. In essence they were Inters in plunger frames with internals best suited to racing and were supplied to selected fancied runners with such helpful items as megaphone exhaust to make them go and conical hubs to stop. All the classic features of the garden-gate Norton were present; scolloped edge tank with cut-away at the rear on the right to clear the carburettor, bolt-through fixing, wrap-round oil tank that extended on the left to occupy the battery space of the more mundane machines, oil filler on the left on an extended neck, large parallel check springs on the front forks and the late type plunger system.

The engine remained the single ohc type and the crankcase lacked the vertical fins of the works bikes, but the cylinder head carried its line of upright fins above the exhaust port that was to characterize the engine from then on and had the full square finning used by the works. The valves and hairpin springs remained exposed.

The new type of plunger suspension was also made an option for the other camshaft models and the ES2 and the former were given some changes to help improve the cambox sealing if not to actually stop the leaks.

And on that note the camshaft models came to a halt, or rather an interlude while the plebeian 16H went to war en-masse helped out by a handful of special Big 4 models.

There is, however, an offshoot of the ohc story that concerns the big banger camshaft engine. This was based on the model 19 dimensions so was also of 597 cc although always known as the 596. Few were built, all intended for sidecar use only, and they came into existence around 1936/37 with Harold Tozer racing an example at Donington in the earlier year. Some were built with iron heads and suitable cycle parts for the ISDT, but most went on to racing circuits. Some came to have the aluminium finned, bronze skull head, and all left Bracebridge Street with a single overhead camshaft. The big engine used the standard crankcase but its height called for a kink in the top tube to enable it to be squeezed in, while the frame itself was heavier than normal. Arthur Horton was the first to use one on a regular basis and later Jack Surtees, father of famous John, had one.

They were few in number but dominated road racing until they were no longer eligible for World Championship events in 1951.

Left **The 1937 works 350 with enclosed valve springs ridden by Jack Williams into 10th place. Hairpins may have suited racing but an Inter without oil leaks would have been nice**

Below **Trying hard at Donington Park in the thirties, driver unknown**

3 Manx and Inter

Immediately after the war road racing faced many problems, among them lack of machines, suitable fuel, circuits, tyres, and travel difficulties. Despite this it struggled back into action and, inevitably, used what was to hand and adapted it as needed. So all the prewar machines with any pretentions to being sporting could and were stripped down and readied for racing.

At first under those circumstances the more exotic ohc machines could not be built at all; and had to be modified to enable them to run on the 'pool' petrol, then the only form available. Compression ratios came down with a bang along with power outputs to try to get engines to run and then keep going.

Despite the problems, interesting work went on behind the scenes. A single ohc engine was fitted out with a water jacket and run by the research department of the IAE. In later years a similar exercise was to be carried out using the dohc engine as a basis to determine the real effect of water cooling on power and stamina. The result was more power that held up for a longer period.

A further insight to the future came in July 1945 when the first postwar races were held at Bangor Castle. The McCandless brothers, Rex and Cromie, had a successful day sharing the same machine powered by a Triumph Tiger 100 engine and this unit was installed at a well

Postwar garden gate, the 1946 Senior Manx Norton with Roadholders

Norton

inclined angle in a frame of their own devising. It had duplex tubes running from the headstock above and below the engine and gearbox to a pivot for a swinging fork carrying the rear wheel. A stressed skin assembly went over this to support the rider and the telescopic rear suspension units, while at the front of the machine was fitted a telescopic fork designed by the brothers. All at a time when girder forks and rigid frames were the norm.

It was 1946 before there was news of the camshaft models. It had been intended that for 1940 the road racing Inter, which had become known as the Manx Grand Prix, would have telescopic front forks as standard and be opened up in the bore to 499 cc, while the touring camshaft models would be given the option of the telescopic fork. These forks would have been similar to the prewar works machines without damping, for it was not until late in the war that hydraulic damping was added and tried out on a 16H.

During 1946 racing machines were built based on the prewar models and were supplied in time to run in that year's Manx Grand Prix. They were sold fully equipped for racing right down to the number plates and in this form were given the simple name of Manx Norton. The International models continued for road use and were fitted out for that purpose, while the CS1 and CJ were not revived.

Both Manx and Inter were built in two engine capacities with a letter 'M' added to the model number to distinguish the pure racing machine.

The famous bolt-through petrol tank with wing-nut filler cap on a well restored 1946 Manx

As always the model 30 was the 500 and the 40 the 350 and engine dimensions remained the traditional 71 × 88 mm and 79 × 100 mm, except for the 30M which was enlarged a little to 79·62 × 100 mm and 498 cc.

The compression ratios were the same for both machine types at 7·33:1 for the 350 and 7·23:1 for the 500 as supplied, but the Manx had more compression plates than the Inter. The latter was provided with one only, 1 mm thick, whose removal lifted the ratios to 7·8 and 7·6 respectively, but the Manx had two plates at 1·5 mm and a further at 0·5 mm so the ratios could be taken in stages up to 9·5 and 9·0 for the two engine sizes.

The Manx engine used much from the prewar works units but retained the single camshaft. The cylinder head was square, about 10 × 9.5 in. and in aluminium cast to a bronze skull, while the cylinder was alloy with an iron sleeve and the crankcase magnesium alloy. For the rest it was the Carroll engine as before with a rev-counter drive from the oil pump and magneto drive. The Inter looked very similar to the Manx although it retained the iron engine and aluminium crankcase but with hairpin valve springs and a downdraught inlet port.

Both engines drove the Norton four-speed gearbox with the Manx fitted with close ratios and no kickstarter, while the Inter had that aid to starting and road ratios. The Inter gearbox also had a new end casting and one-piece outer cover which streamlined the assembly without changing the manner in which it worked. The new cover enclosed the positive stop mechanism as before but extended down to encom-

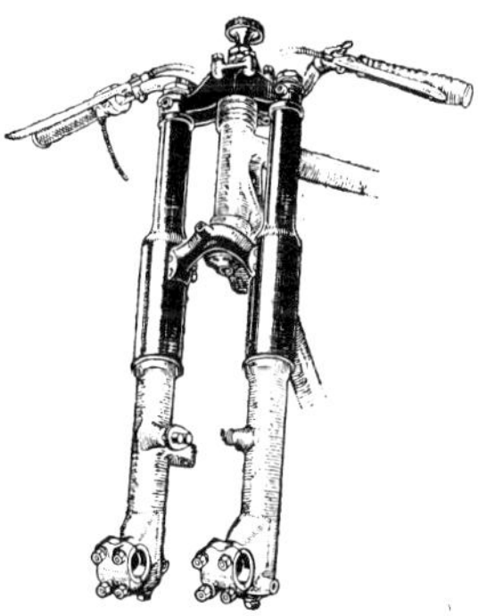

Left **The offset axle forks used by the works in 1949 and later by Eric Oliver**

Below **Special Amal used on some works machines in 1949**

Right **The works rear suspension with added hydraulic dampers as used from 1947**

Right below **Twin leading shoe brake with internal linkage adopted in 1948 front and rear**

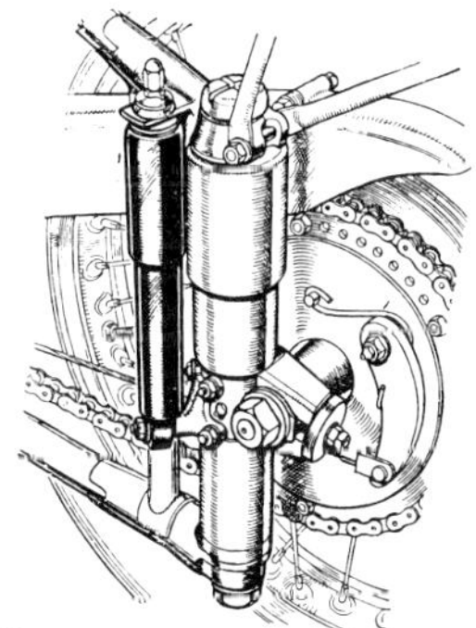

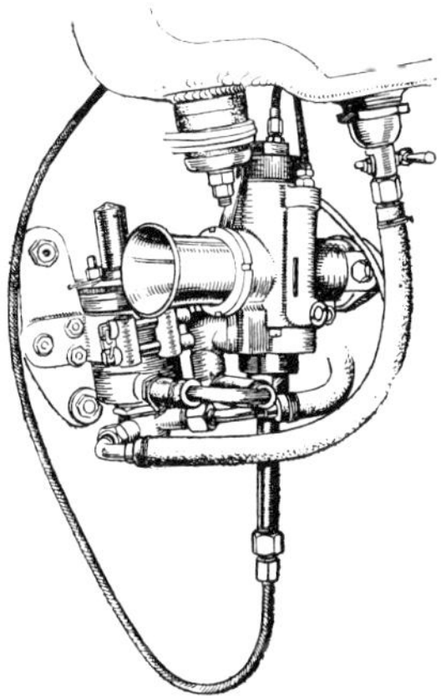

pass the clutch worm and the kickstart spindle. The end casting continued to carry the filler plug and other items and was just altered to mate with the cover. Access to the clutch worm and cable end was via a hole in the cover blanked off by a circular pressed steel cap at first. This was not very satisfactory and by the middle of 1947 it had been replaced by a small cover plate secured with two screws.

The Inter clutch was standard Norton as was the chaincase, although this did have a black and chromium finish, but the Manx came with an open chainguard and a light, racing clutch with three plates and Ferodo bonded asbestos linings.

The frame of both Manx and Inter models was the full cradle type with plunger rear suspension, known as the 'garden gate', while at the front went the hydraulically damped Roadholder telescopic front forks. Construction of these was simple and straighforward with each fork leg fitting into a taper in the top crown and being clamped at the lower. A cap nut pulled the leg into the upper taper.

The main fork tubes were in chrome-molybdenum for the Manx models and Reynolds A quality for the Inters, both however of the same 1·344 in. diameter. The lower sliders were in alloy with an oil seal working on the fork tubes and above them went two springs in series, the main one which deflected at 40 to 45 lb per in. and a short buffer at the top with a 200 lb per in. rating. Progressive damping was achieved with a restrictor plug at the bottom of each fork leg through which passed a plunger attached to the slider. This had a parallel length around the suspension middle position and tapers either side to restrict the movement of the damping oil as the fork deflected more. Further larger parallel sections acted as hydraulic buffer stops in either direction.

On the Manx the forks had different top shrouds to the Inter as it was only necessary to

Artie Bell at Nutts Corner in the 1949 Ulster in which he was 2nd on his 500

4

support the front racing plate which carried a fine mesh flyscreen at its top. The bars and rev-counter were mounted on the top crown along with the André friction steering damper. The Inter shrouds carried the headlamp with its light switch and ammeter in a small panel, while the speedometer was flush-mounted into a neat panel which bridged the fork tubes.

The Manx had an 8 in. diameter conical front hub complete with air scoop on the alloy backplate but the Inter managed with the standard 7 in. units used by the rest of the range. Wheels were not standard, however, with the Inter running on the same size of ribbed Dunlop at the front as the Manx, although the latter was on racing rubber not the usual road compound. At the rear went triple-stud Dunlops with the Inters and 40M sharing the size and the 30M one section wider. In all cases the wheel diameter was larger than that used by the less exalted ohv and side valve models.

The Manx was equipped solely for racing so many of its details were only suited to that purpose. For the engine this meant a racing BTH magneto was fitted to fire the mixture which was supplied by a remote needle racing Amal and exhausted out of an open pipe with megaphone. The gearbox retained the old type end casting as before the war. The fuel tank was still the bolt-through type and had the rear cut-away on the right to ensure adequate clearance around the carburettor bellmouth. It also had a quick action, large diameter, filler cap and a chinpad, while the oil tank was the wrap-round type with the filler, again quick action and large, on the left to suit the pits in the TT. A rear mudguard pad was provided as was a centre stand, while the rear mudguard stay could be used as a lifting handle.

The International models had slightly smaller tanks, although still larger than standard, and the oil filler was on the right while the battery went on the left which reduced the tank size. Normal wing nut filler caps were fitted. The rest of the Inter equipment was stock road model with full electrics, mudguards, saddle and stands. It was also fitted with the tubular silencer which did not like living with the performance cams used. This was usual Inter practice and at that time of no great concern for most prospective owners who were ordering models for the new Clubman TT race to be held in 1947. In this, machines had to be available and sold as road models, which ruled out the Manx, but could be raced minus silencer and electrical equipment. Megaphones were not allowed but the open pipe suited the Inter down to the ground. A centre stand was fitted but not one at the rear.

Two events that were to have a considerable effect on the fortunes of the Norton firm and their racing successes occurred at the end of 1946. The first was the return of Joe Craig from AMC to take up the post of technical director, and the second the banning of superchargers combined with a restriction to the use of pump fuel for road racing.

Nortons did not have a blower but that ban removed the Gilera threat for a year or two and had repercussions for the AJS Porcupine. However, the fuel which was then pool petrol of around 72 octane gave the big singles violent detonation problems and by April Joe Craig was talking of ratios below the 7·2 and 7·6 used in the 1946 Manx. Pool was apparently even more vicious than he thought but while Joe reminisced about the 10·5 and 11·1 used in 1938, he just might have been sowing seeds of doubt in the minds of the opposition.

Whatever the problems may have been, by May things looked better and in June Norton won the Senior TT once more. The machines used by the three works riders were fitted with the 1938 engines even down to the remote needle carburettors which were renovated by Amal. Problems in that area arose from the loss of all drawings and patterns in a fire so that new units were not available. Engine dimensions of the 350 went to 73·3 × 82·5 mm, ones that had been tried on Frith's engine in 1939 and were to

Left **A National Motor Museum Manx—a garden gate in all its glory, fully restored. Shot at Earls Court**

Below **The twin cam Manx engine installed but minus petrol and oil tanks**

remain for a while. The cycle parts were brought up to date with Roadholder front forks and hydraulic dampers added to the rear suspension.

The performance was down a little from the 125 mph of the 1938 works 500 to about 118 mph while acceleration also suffered. Handling and the expected Norton steering were all up to the mark and helped Harold Daniell turn in an 82·81 mph race average, although he thought the prewar machine steered better. Due to his preference for harder springing he raced without the rear dampers and thought the front end jumped about a bit.

In the middle of 1947 both *Motor Cycling* and *The Motor Cycle* had 490 cc Inters on test and both were supplied with strictly non-standard silencers of the old Brooklands form with fishtail. For the *'Green Un'* Charlie Markham squeezed a flash reading of 97 mph from what was effectively a 1932 engine plus a flying quarter at 94·71 mph

Bill Boddice followed by Jackie Beeton at Cadwell Park. The chairs are typical of the period and based on grass track practice

and readings of 91 and 83 mph in the gears. The *'Blue Un'* either had a less well prepared model, a bigger rider, or did not try as hard for their figures were 86, 80 and 69 mph, although both magazines returned standing quarters in similar times.

There were no alterations to the ohc models for 1948 but when TT time came round there were changes for the works motors. The engines had a number of new features and for the first time the 350 and 500 had a common crankcase. This item had a larger capacity oil sump and was stiffened up at the rear, while along its base were the four studs that were to characterize the engine from then on. Internally the pistons were improved while the cylinder head carried deeper finning on the outside. The hairpin valve springs were given more freedom and thus less stress by

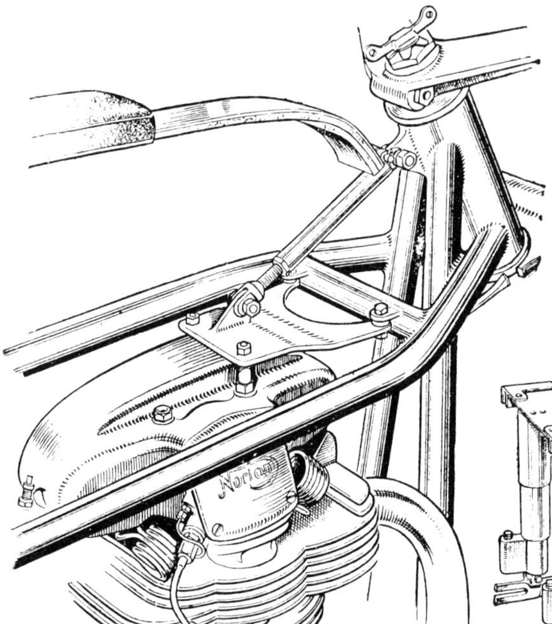

Above **Featherbed prelude, a picture taken in October 1944 which shows the forward thinking of Rex McCandless with his Triumph powered chassis**

Left **The featherbed frame in its first form with bolt on subframe and remote reservoir units. Detail shows head steady and tank strap**

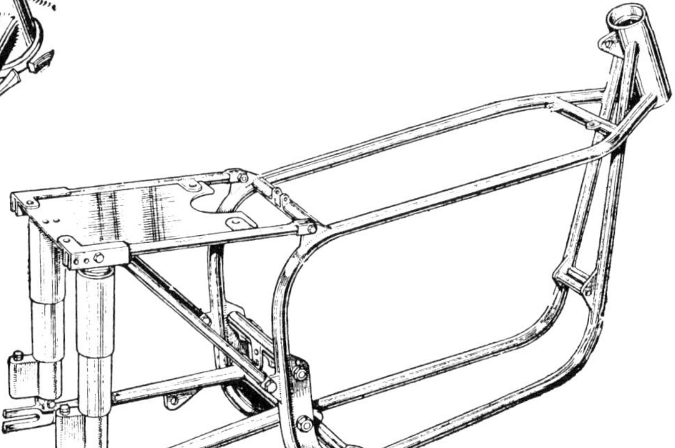

fitting their lower legs into slots instead of holes. Less obvious was the adoption of Wellworthy Alfin cylinder barrels which gave much improved thermal conductivity and whose use was kept quiet until 1950.

The engine rear mounting was raised to increase the support it gave in the frame and this in turn led to the magneto platform being raised an inch and in turn the advance cable operation being modified as the magneto came too close to the oil tank for the old layout. The oil tank itself had the breather tower moved to a central fore and aft position to make sure nothing came out under the hardest braking.

A new type top feed float chamber was used by several makes that year but the one on the works Norton was larger than standard with a flexible mounting to the mixing chamber and a flexible feed pipe. The adoption of the new type was to combat float flooding and provide a steadier fuel supply. The float chamber went on the right side of the machine as the area to the left of the carburettor was filled by the oil tank and that in turn had an extended neck for the quick action filler.

The frame was as before with Newton hydraulic dampers helping the rear suspension but the height of the whole machine was reduced by fitting a 19 in. rear wheel and shortening the front forks to maintain the angle of rake and the trail. Along the same line of thought, the steering damper was moved to the underside of the bottom crown only leaving the knob at the top, while vision past it was improved by fitting the rev-counter vertical in a bracket held by the right top fork nut.

On the transmission side a four-plate clutch was fitted and, while it remained out in the draught, the major part of the primary chain run was enclosed with an outer cover. This together with a guard over the top run of the rear chain was all part of a move to keep oil away from the rear tyre.

A smaller diameter exhaust pipe went on the 350 and, after experiments in practice, twin leading-shoe brakes were adopted front and rear. This was done by modifying the back plate so that while each shoe retained its separate pivot post it also had its own cam, and these were linked by an internal connection so that the single outside arm remained the only lever to be worked.

The production camshaft machines carried on into 1949 with little change but the Inters had the brakes modified and the rear lamp positioned central above the number plate. The brake change was introduced for the new road twin and used on all road models and was simply a tie-plate to link the two shoe pivots as had been done by private owners for a good many years.

1948 had seen what was to be the last of the bigger banger single camshaft jobs and the only one built postwar and fitted into a plunger frame. It was a special job for Cyril Quantrill, then sports editor with *Motor Cycling*, who wanted something as big as his model 19 but with more and sturdier horses to pull a sidecar along the fast roads of Europe at a demanding schedule. Gilbert Smith gave the idea his approval and the works managed to find all the special parts they needed in the spares bins. In due course Cyril collected it and it spent the next few years covering race meetings and ISDT events very rapidly.

There was one more 596 to come, but that was an engine only job. When the world championships were announced for 1949 things did not look too rosy for the Norton race team. They knew the Velocette had the legs of them in the 350 class, while the new AJS was right on their heels as they were using what were effectively de-tuned prewar machines. In the 500 class they had to contend with Gilera, AJS and an occasional Velocette, and in the end their only solo success that year was in the Senior TT and that only because the AJS stopped.

That year the sidecar class ran at 600 cc so the old 596 camshaft engine was eligible although with the opposition on Gilera 4s and BMWs that

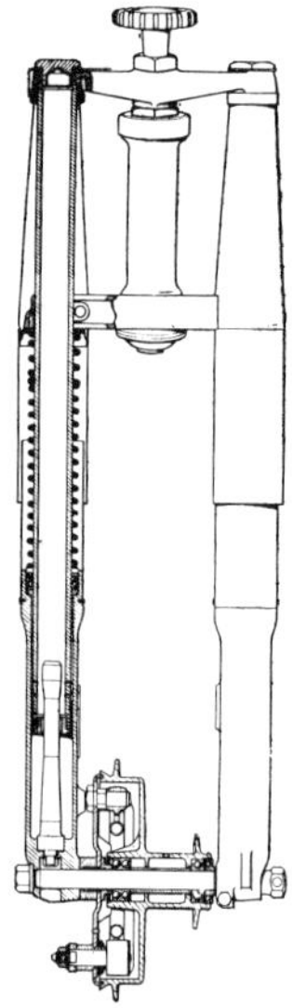

Above left **First of the many, Duke's 500 as seen at Blandford in April 1950**

Above right **The Roadholder forks and one piece front hub**

class also looked unlikely to yield much to Norton. However, they held a very real ace in that Eric Oliver had been campaigning a big banger chair for some time and was a genius on three wheels. He, with passenger Denis Jenkinson, stood head and shoulders above the rest and his artistry could more than compensate for any lack of power.

So he was a man to back and Norton did this by building a twin overhead camshaft big banger engine for him. This had an all special top end for the extra bore diameter spread the valves and meant a one-off cambox to get the cams out enough to be over them. The outfit ran with rigid frame and girder forks and Norton's faith was rewarded when Oliver easily took the championship for the first, but not the last time.

The most noticeable change to the works machines for 1949 was to the front fork which altered completely to mount the legs roughly in line with the steering column with the wheel spindle ahead of them. The front hub was also new with the mass more in the centre and an improved brake plate. Its spindle was large and hollow to reduce weight and secured by split clamps each held by four fixings. Thus the spindle played a considerable part in preventing the fork legs twisting.

A detail improvement was the addition of a spring under the steering damper knob, while the petrol tank vent was moved off the filler cap. The oil tank cap was angled a little to aid filling and the float chamber changed to the standard size top feed. It was moved to the left of the carburettor where it was more on the centre line of the machine and mounted from the seat tube. On one or two machines a special Amal was fitted that had a second cable running from below into the jet cap nut. This was connected to a needle which it pulled down as the slide rose to give a clear bore at full throttle as on the remote needle type, but by reducing the distance between the needle jet and choke tube gave more precise mixture control.

For the faster private owners in the 1949 TT there was an important change to the engine of

Right **Typical early 1950s Norton advert highlighting their racing successes**

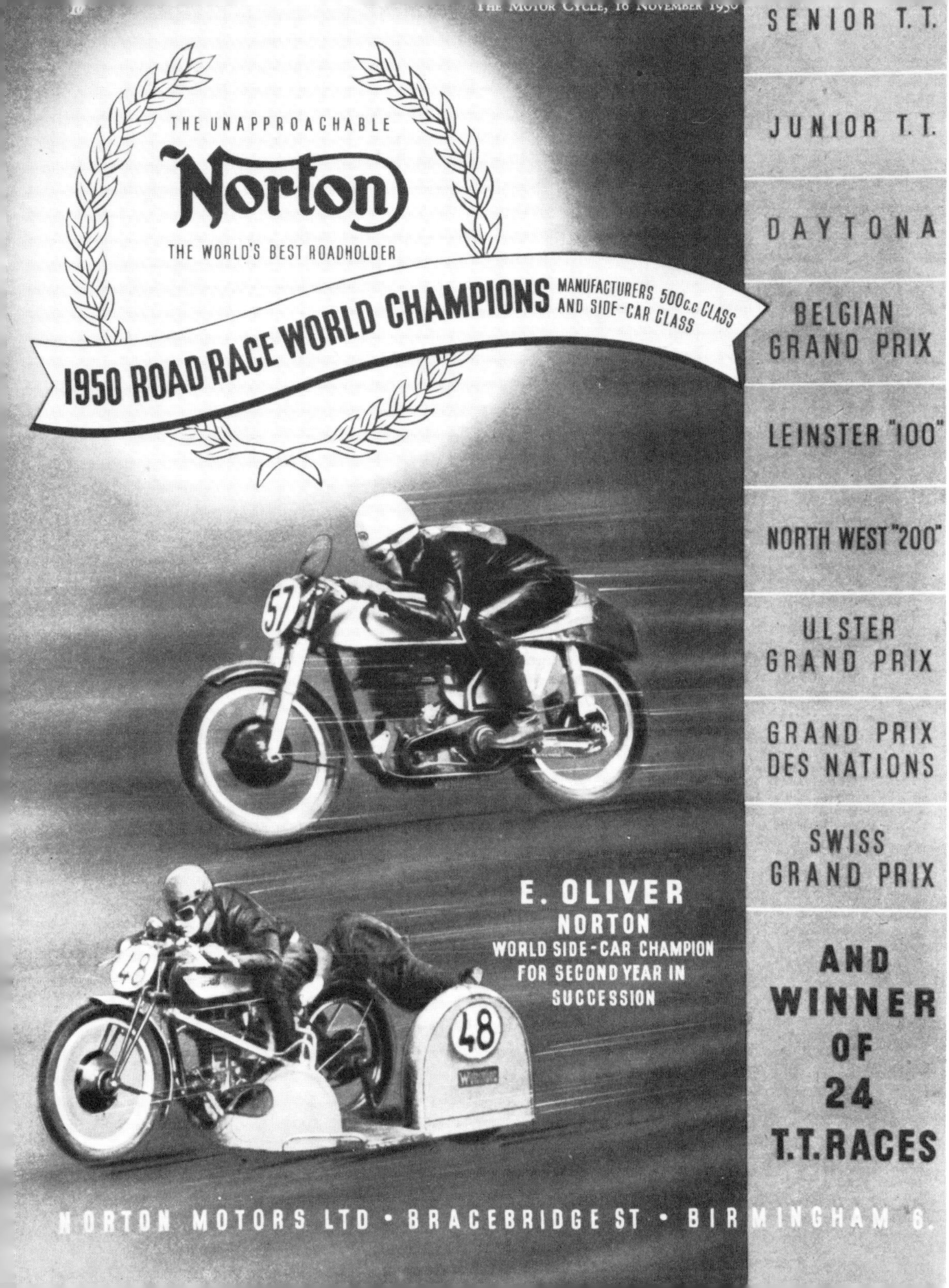
THE MOTOR CYCLE, 16 NOVEMBER 1950
THE UNAPPROACHABLE
Norton
THE WORLD'S BEST ROADHOLDER
1950 ROAD RACE WORLD CHAMPIONS
MANUFACTURERS 500c.c CLASS AND SIDE-CAR CLASS
SENIOR T.T.
JUNIOR T.T.
DAYTONA
BELGIAN GRAND PRIX
LEINSTER "100"
NORTH WEST "200"
ULSTER GRAND PRIX
GRAND PRIX DES NATIONS
SWISS GRAND PRIX
AND WINNER OF 24 T.T.RACES
57
48
E. OLIVER
NORTON
WORLD SIDE-CAR CHAMPION FOR SECOND YEAR IN SUCCESSION
NORTON MOTORS LTD • BRACEBRIDGE ST • BIRMINGHAM 6.

their mounts which, as was common practice, were delivered to them only shortly before the event. It was the adoption of twin overhead camshafts and, while the new cambox looked as if it would fit straight onto an earlier engine, in fact a new cylinder head was also needed along with frame and petrol tank changes to accommodate the new parts.

The new cambox was driven just as before so the familiar angled cover plate at the top of the drive shaft was retained. The cambox itself was in two parts with an outrigger plate which bolted into the main section. This section had the vertical shaft drive housing bolted to it as before and extended across the head enough to encompass two bushes placed below the cams at an angle for the hollow pushers which took the cam side thrust and moved the valves. A total of five shafts ran in the cambox with the centre one driven by the bevel through the usual peg and hole vernier which allowed the whole cambox timing to be varied. This shaft ran in journal bearings with a pinion keyed to it which drove a larger intermediate gear to each side. Each turned on caged double row rollers running on a spindle fixed into the main section of the cambox and held in position by a nut, while the other end was supported by the outrigger plate which also carried the bearing for the central shaft.

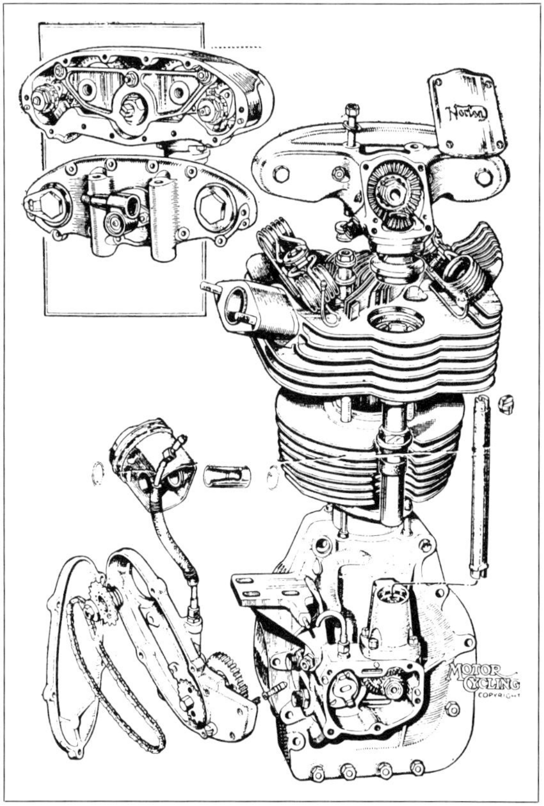

The 1950 works 500 with diagonal head fins and integral top bevel housing

The camshafts lay at the ends of the cambox, each driven by an intermediate gear, and in journal bearings located in the main cambox section and its left side cover. Apart from the bearings, this cover simply enclosed the gear train and the outer bearing races were set into it by large cap bolts with external locking washers, and these plus various shims determined the camshaft endfloat and pusher to bearing clearance. The pushers were also lapped into their bushes so assembly of the cambox was a skilled and tedious job. The cams were separate from their shafts and repeated the peg and hole vernier so both could be set for optimum valve timing.

The lucky owners of the new engines also had light alloy tanks with the oil tank fixed in the same bolt-through style as the petrol tank. Alloy wheel rims were also fitted.

The standard Manx and Inter models were continued without change when the 1950 range was announced, but for the works it was to be the year when the supposedly outmoded single gained another lease of life, two in fact for it was the year of the featherbed frame and the emergence of Geoff Duke at the forefront of road racing. Duke had already made his mark very firmly in the Clubmans and Manx races so was an expected selection for the 1950 works team. His

Right **Miss Teresa Wallach in London following a 30,000 mile tour of the USA and Canada in 1950. In 1939 she won a Brooklands Gold Star on a 350**

Norton
THE PETRIFIED FOREST
THE PAINTED DESERT
THE GREAT SALT LAKE
UTAH
DENVER
WISCONSIN
THE EVERGLADES

G. Beresford

The 1951 Manx Norton as sold to the private owner with featherbed frame, long stroke engine and single leading shoe brakes

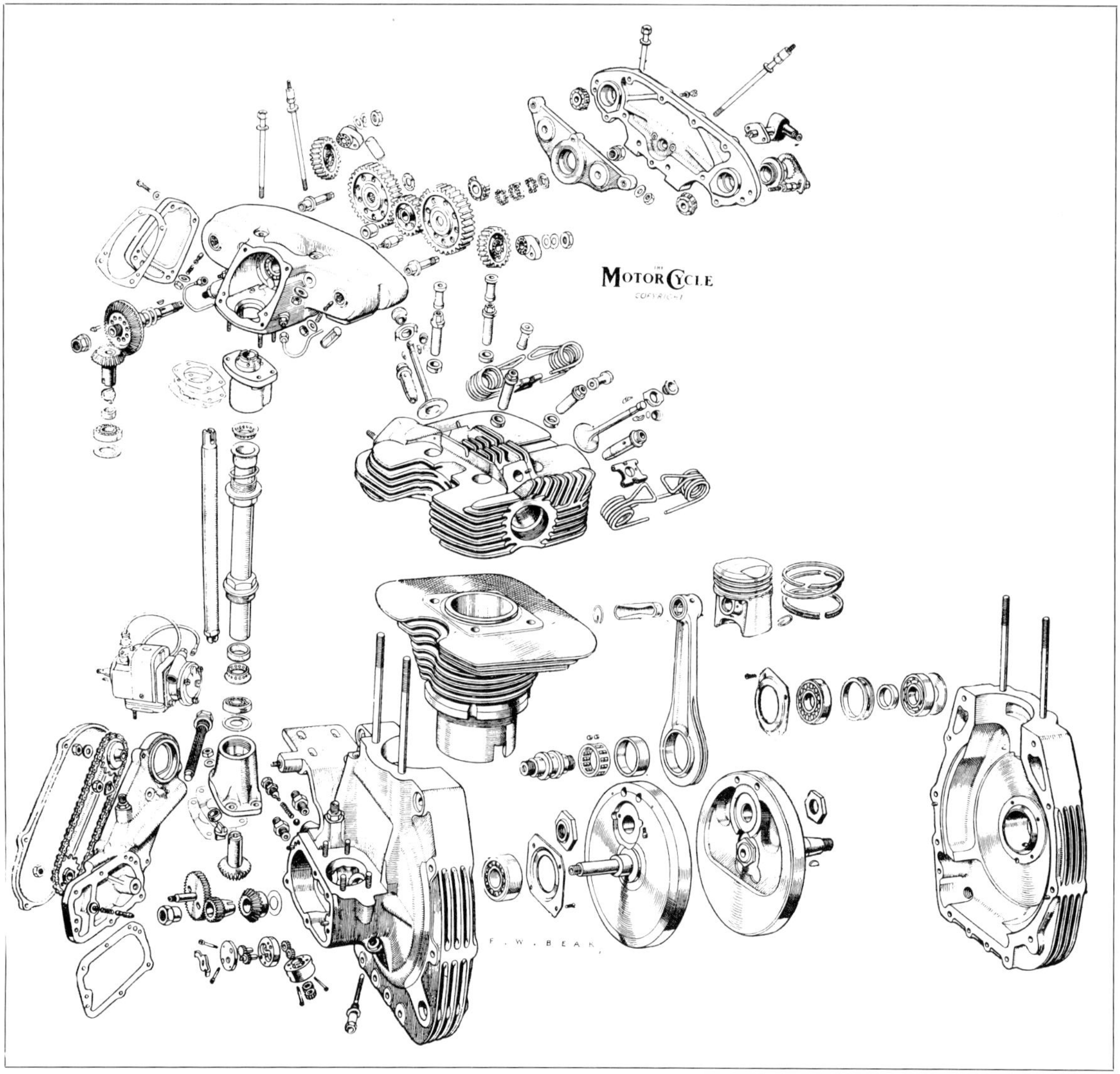

Exploded view of the 1952 production Manx engine, not an easy one to assemble well

totally dedicated riding style suited the new frame and in no time he stood head and shoulders above his contemporaries.

The new 500 works Norton made its debut at Blandford Camp circuit in Dorset on Saturday 29 April 1950, carrying race number 43 and ridden by Duke. When Joe Craig pulled the enveloping cover from the machine a new era began and totally new standards of road-holding, steering, and rider comfort were set. The name of 'featherbed' was coined by Harold Daniell after his first ride.

The new frame was a fully duplex, swinging fork type built up from two main loops of tubing. Each loop ran from the bottom of the headstock, itself a tube and not a forging, along under the

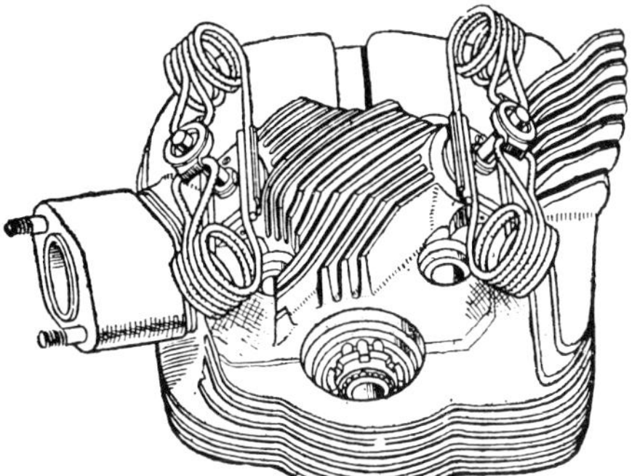

The 1950 works head with diagonal fins on top and vertical above exhaust port

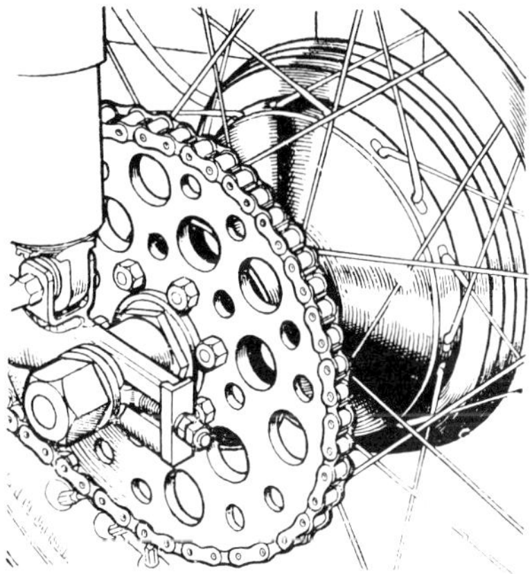

Rear hub with brake drum remote from rear sprocket, tried in 1951 but not raced until 1952

petrol tank, turned down behind the oil tank and gearbox, turned again forward under the box and engine to the final bend and up to the top of the headstock. The down tubes passed between the tank tubes, and the two loops were joined by cross tubes behind the headstock, in front of the crankcase and under the back of the gearbox. All these tubes were sif-bronzed together and gusset plates were added to each loop at the lower part of the seat tube to provide a pivot point for the rear fork and to brace the frame. At the top of the gussets a further cross tube was bolted in place.

A light sub-frame was bolted to the main assembly to support the seat and the tops of the rear suspension legs. These were positioned vertically and under the covers were both springs and hydraulic damping, the oil for the latter in reservoirs built into the back of the lower part of each leg. The rear fork was built up from tubes in the same manner as the main frame and pivoted on Silentbloc bushes.

At the front of the new frame went the Roadholder telescopic forks minus their top shrouds and carrying clip-on bars in place of the old one-piece bar clamped to the top crown. Further moves to reduce frontal area were the recessing of the rev-counter between the fork legs and well down behind the front racing plate, which was slightly faired at the edges and carried a mesh flyscreen.

The height reduction was helped by the use of a 19 in. front wheel fitted with a 3.00 in. section cover, and the whole appearance looked far more tucked-in than had ever been the case with the garden gate model. The low ride-height was helped by the use of a 3.25 × 19 in. tyre at the back and the new style fuel tank allowed by the new frame. This had no tunnel so was really a simple light alloy box sat on rubber pads on the top tubes. In fact it had knee recesses at the rear and was moulded to suit a crouched rider, but as a simple box it immediately removed many of the problems of the older type. It also held $5\frac{3}{4}$ gallons so was very suited to long races. Attachment was simply by a strap and buckle which was quick and easy to remove.

The 8 pint oil tank sat between the frame tubes above the gearbox and was cut-away on the right to clear the carburettor bellmouth. It incorporated a small platform welded to its front side on which sat the float chamber. Rather than continue the practice of a filler cap on one side or the other, which could be awkward for some circuits, it was dead central in the top of the tank. Access was gained by a cut-out in the front of the one-piece seat adopted in place of the old style saddle and pad. This seat had a humped back to which a small fairing was attached and this also acted as the rear number plate.

Behind the new frame lay eight years of work for it had been 1942 when Rex McCandless and Artie Bell had first had thoughts of ways to make the racing 500 more controllable. The results had first been seen as the Triumph-engined special built by 1944 and raced successfully at that meeting at Bangor in 1945. Later had come conversion sets for many makes, and fame for Artie Bell as a Norton works rider. With this link the factory began to co-operate with McCandless and, after many tests, the prototype frame was taken to Birmingham in a form able to take the camshaft engines, and all the others in the Norton range.

During the early months of 1950 the new machine went to the Isle of Man and was tried on some of the mountain roads after which it was given a 400-mile thrash round Montlhèry plus 600 miles at MIRA, which failed to bring out any flaws. Then the first of the season's race machines was built and from then on it was the debut at Blandford where Duke won easily, followed by success in two major Irish events and then total domination of both Junior and Senior TTs with Duke breaking the long-standing Daniell lap record.

It was not only the frame that played a part in this success, for the engines were revised as well. They were still the same Carroll design of 1930 and still retained dimensions of 73·32 × 82·5 mm for 348 cc and 82 × 94·3 mm for 498 cc as used for some time. Of real help in improving both power and reliability was a change in the racing rules that allowed 80 octane fuel to be used and this made things much easier for the big bore single.

The most noticeable change to the engine was to the cylinder head which had the upper bevel housing cast into it with the horizontal fins running around the housing. The fins were also closer pitched and above the combustion chamber were arranged diagonally under the

For those without a featherbed, a Roy Cunliffe conversion to a garden gate. Thanks to Phil Heath we also know that the engine is an Excelsior Manxman

The rotary valve engine built in 1952 and based on a works motor. Combined valve drive and oil pump in lower bevel box

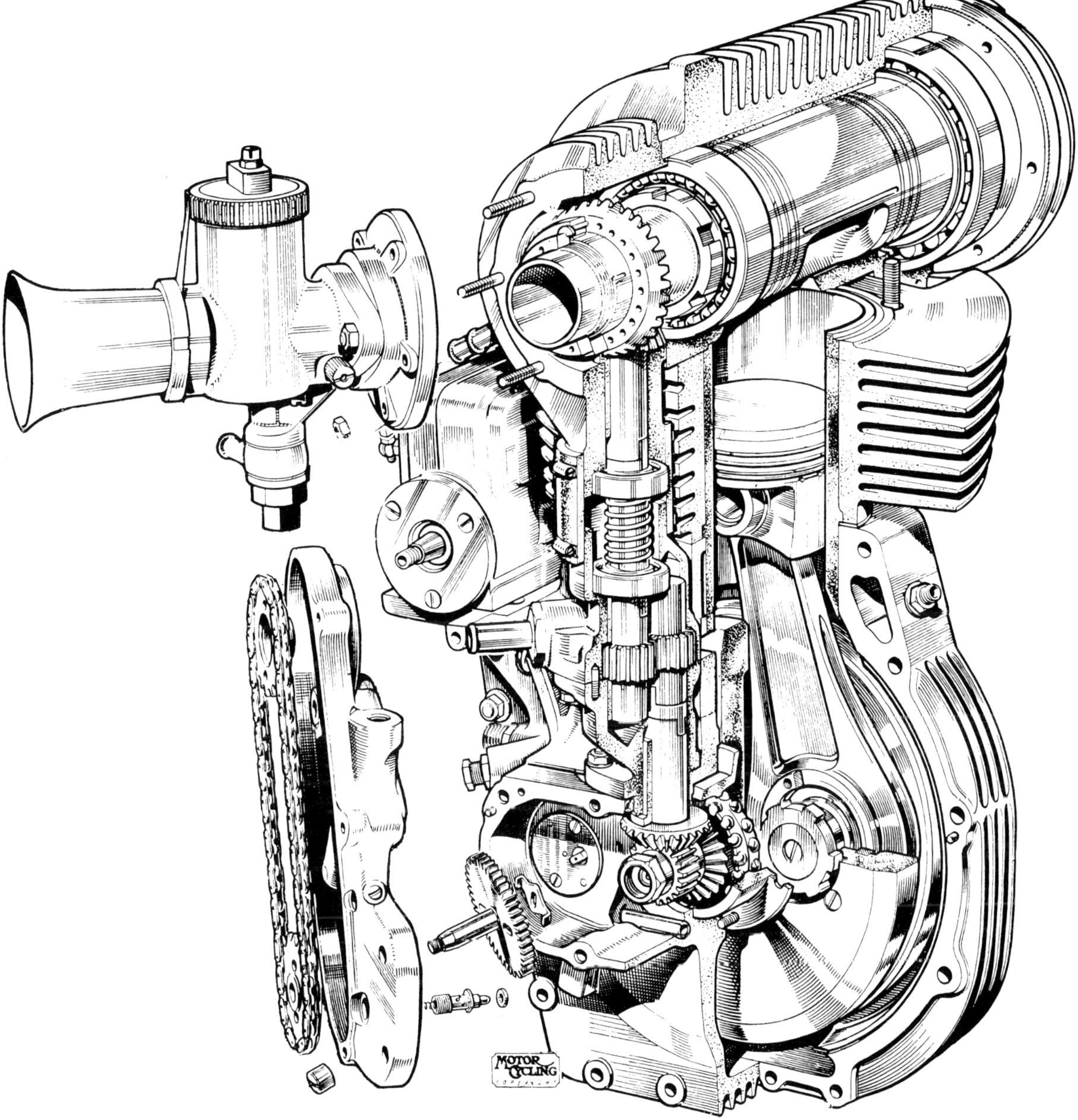

Unusual, maybe unique, chain driven camshaft engine built by E Jackson of York and based on 1936 model 30

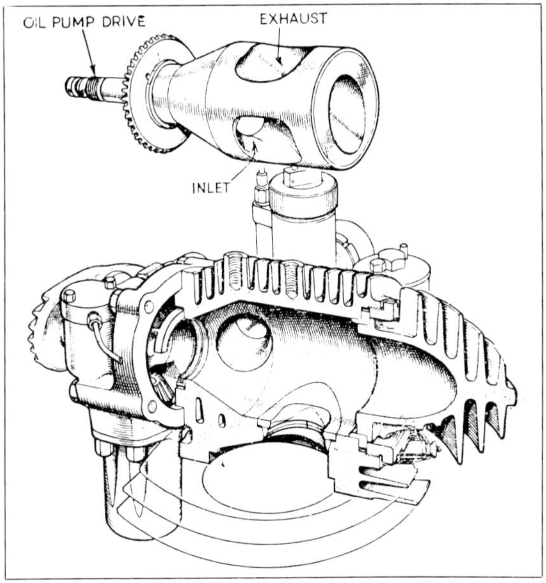

Line drawing of head and rotary valve conversion made by Brown brothers in 1947

cambox so that the cooling air flowed from left to right and emerged from behind the bevel housing. The head itself was no longer bi-metal but cast in aluminium with iron valve seat inserts, while the barrel continued to be a Wellworthy Alfin. The cambox had the fixing bosses on the cover extended to make its mounting to the head more rigid, and the rev-counter drive was taken from the left end of its centre shaft. This removed the drive box from the magneto cover, where it could drag on the ground and be knocked off, and reduced the drive cable length.

Above the cambox and attached to its two forward fixing bolts went an engine steady plate which ran forward to bolt to the frame cross tube just behind the headstock. The plate also carried a lug to which an adjustable stay was attached and this ran to the top of the headstock to a lug which also acted as the hinge point for the tank strap. The stay was typical Norton with a massive threaded adjustor and matched by the weighty steady plate. They didn't break though!

Lower down in the engine the crankcase external ribs were extended out to the fixing on the drive side and a gauze filter was added to the bottom bevel chamber, the filter being attached to a large hexagon plug inserted in the back of the chamber just under the magneto platform.

At first the engine was fitted with a modified remote needle Amal but for the TT used the prototype GP type developed from it to give the best points of both RN and TT models. On the outlet side a shorter exhaust pipe and smaller megaphone were used, indicative of higher engine speeds, and the power band was much wider than that of the earlier motors with their very large trumpets.

The engine drove the usual Norton clutch with its shock absorber centre by an open primary chain with just a guard over the runs and minus the covers of the previous year. The gearbox was mounted between the rear engine plates which extended round and behind it to the two rear fixing points. The box was pivoted on its lower through-bolt and the primary chain adjusted by moving the upper one with a screw. The box shell used was that from the twin with simple lugs top and bottom and the positive stop mechanism at the front of the mainshaft instead of above it. The gearbox internals were as before with the

Right **Ray Amm on the grid for the 1953 North-West 200 with the Silver Fish**

MASTER
IN OILS
21

camplate at the front of the box but with its detent spring now on the underside. It was turned by an internal quadrant and this linked directly to the change mechanism. In this form the box became known as the 'laid down' type and for the racing application the box pedal shaft was linked to the rider's pedal, which pivoted on his footrest, by an external linkage.

The whole engine and gearbox assembly was fitted to the frame as a unit with two small triangular plates at the front and using the rear engine plates extended back to the rear. The result was a machine with everything tucked in and neat and tidy. Even the control lever pivots were welded onto the handlebars.

After their success in the TT things went a little awry for the works team but they were unlucky not to win a solo rider's title, although they did collect the 500 cc machine one, and Eric Oliver once more took the chair class by winning all the races so that gave Norton a second maker's title.

While the works solo results were none too satisfactory, there was no doubt in anyone's mind as to how good the featherbed was and private owners were delighted when it was announced that the 1951 Manx machines would have the new frame and be available from March.

Before they appeared there was news of a new outfit for Eric Oliver using the featherbed frame with modifications to take the sidecar pick-up points. The offset forks from the 1949 works models were fitted along with the matching hub with its large diameter, hollow spindle. The front brake was from the early 1948 works machines and had an outwardly dished spoke rim on the near side. An all-welded Watsonian chair went with the machine, which was powered by a 500 cc

Ray Amm doing the TT double in 1953, here seen on his Senior works mount at Hillberry

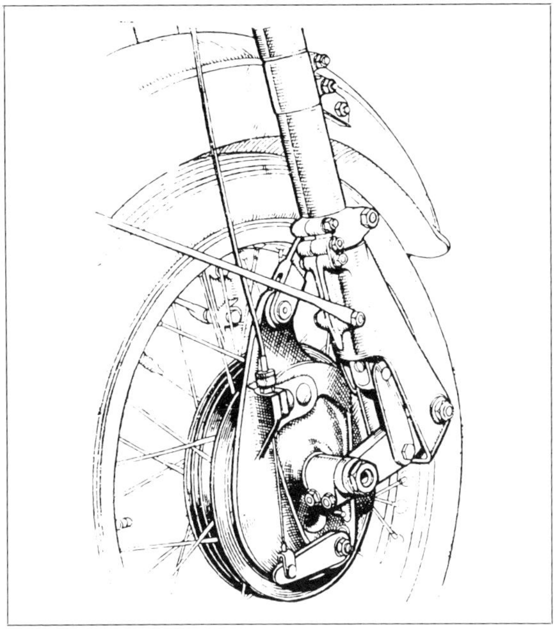

The trailing link front fork tried by Amm at Floreffe and in TT practice in 1953

engine due to the change in the rules for that year. The normal rear suspension was retained, although at the time most thought a rigid frame best for a chair.

The Manx models proved to be very similar to the 1950 works machines with the same frame but with the rear suspension controlled by units lacking the remote reservoirs. The forks had longer springs fitted inside the main tubes and the machines ran on 19 in. wheels with alloy rims and 3·00 in. front and 3·25 in. rear racing tyres. Brakes were single leading-shoe with the rear backplate modified to suit the swinging fork frame.

The engine was much as before with very deep crankcase moulded on the lines of the 1950 works ones, and inside it turned the 1950 flywheels. The gauze filter was added into the back of the bottom bevel chamber and externally the exhaust port was modified to allow the

The first Streamliner used by Eric Oliver in 1953. A kneeler much as the Silver Fish with floor tank by rear wheel

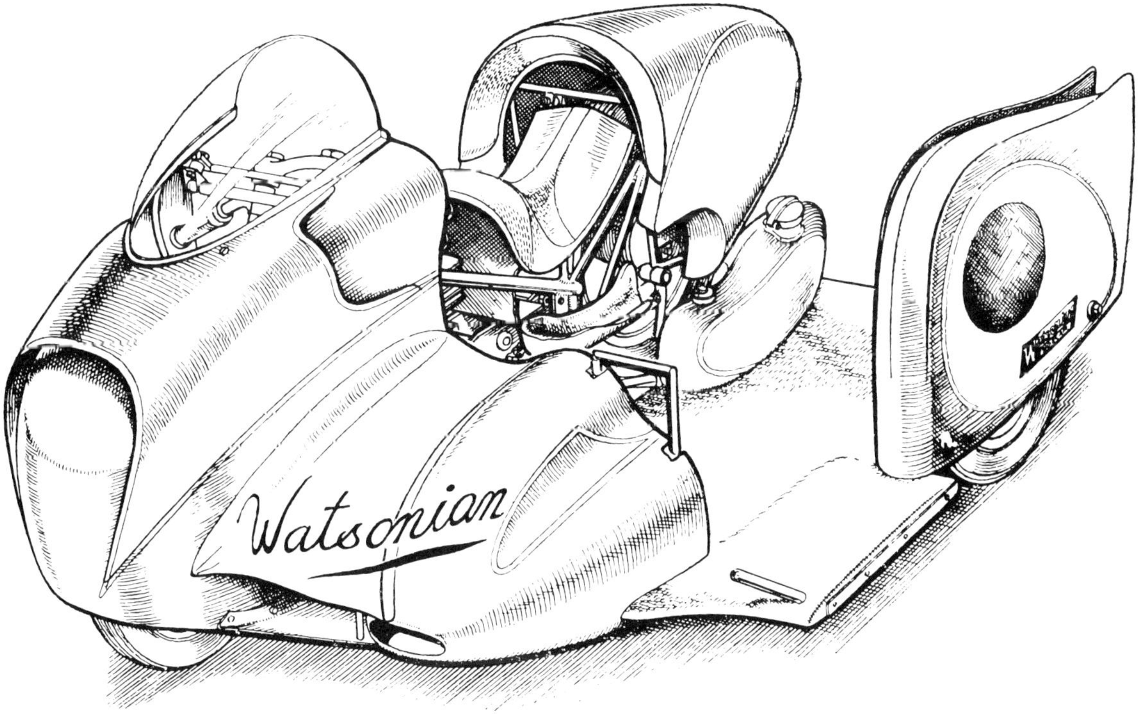

pipe to run just outside the right frame tube. The rev-counter drive was taken from the centre of the cambox. A laid down gearbox was used and normal mudguards to conform to the then new regulations, while there was no seat tail as used on the first works machines.

On the works machines there were more far reaching alterations. The engine dimensions changed once more to shorten the stroke, with the Junior model reverting to the prewar 75·9 × 77 mm and 348 cc while the other went to 84 × 90 mm and 499 cc. The cylinder heads were revised to provide machined faces for the cambox to attach to in place of the less satisfactory distance pieces and more fins appeared around the sparking plug. Outside the engine a GP carburettor was used fed from a float chamber suspended from a rubber mounting which was fixed to a plate bolted across the frame top tubes. This set a pattern to be copied by all.

On the exhaust side the pipe was swept back to pass across the magneto drive cover which reduced its length. While the 500 continued with a large megaphone, the 350 was fitted with a much shorter one with a reverse cone. The 350 also had its inlet tract inclined more than the bigger engine.

The gearbox was slightly altered to improve the entry angle of the clutch cable and this connected to a smaller clutch worm fitted with a ball thrust race.

The frame was as 1950 but built from lighter gauge tubing and the rear legs no longer had the external reservoirs. At the front went the forks with longer internal springs, while regulation size rubber mounted mudguards were fitted. The seat tail was dispensed with.

It was intended to use a rear hub with the brake on the right away from the sprocket but this was not finished so the 1950 type was used. The rear wheel went down to 18 in. with 3·50 in. section tyre, while the front stayed at 3·00 × 19 in. Both chains were lubricated by dripfeeds with

The featherbed International as built in 1953/54 using the 88 twin chassis assembly

the primary reservoir just above the chainguard and the secondary one arranged by utilising the left rear fork leg.

The works team, led by Geoff Duke, had a very successful season with Duke taking the world title in both classes. They were backed up by Oliver who won his third sidecar title in a row with Lorenzo Dobelli in the chair for the second successive season.

For 1952 the standard Manx and International models continued as they were although the first continued to be improved on data fed back from the works models. However, at that time the Manx still retained the rather long stroke engine dimensions and the cylinder head lacked the diagonal top fins of the works engines, while the cambox still sat on distance pieces. In essence the lower half could trace its origins back to the

first Carroll engine very easily despite the passage of over two decades. The crankcase was much deeper and heavily finned, while the drive side mains went into a top-hat housing and were held by an inner plate fixed by four screws. Details altered but the conception rolled on.

The works 500 had another dimension change to the virtually square figures of 85·93 × 86 mm and 499 cc for 1952. On both engine sizes the cylinder barrel fins were extended to surround the vertical drive shaft tube whose material was changed to light alloy and fixings modified. The inlet trumpet of the carburettor was altered in shape and magnetos changed to the rotating magnet-type.

On the chassis side the sub-frame was welded to the main loops along with the upper cross tube and the seat mounting welded onto the sub-frame. The rear spring units were angled forward a little and their construction modified to cast alloy spring boxes. The rear fork bushes were changed to sintered bronze and the rear brake moved to the right side, remote from the

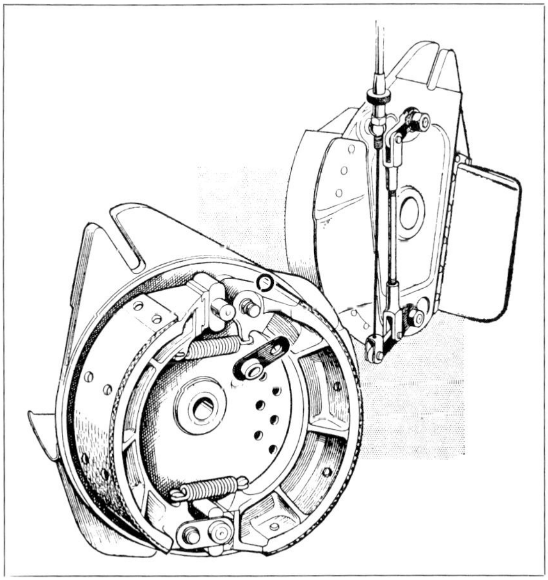

The twin leading shoe Manx front brake with external link and twin, braced pivots

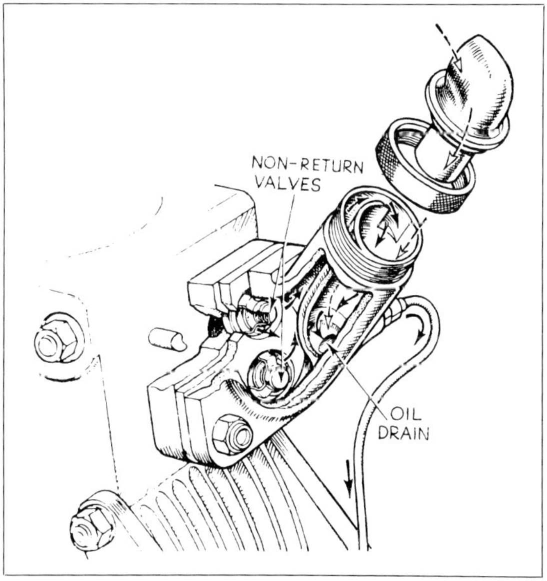

1954 works breather fitted to front of crankcase with oil drain

17

Above **The 1954 works 350 with outside flywheel and many other special features**

Top **Ray Amm at the Circuit di Faenza in 1954**

Left **John Surtees well down on his 350 at Cadwell Park in 1954**

alloy sprocket, and cable operated. The 500 ran on a 4·00 × 16 in. rear tyre mounted on a WM3 rim, but the 350 retained its 3·50 × 18 in. from the year before.

On the three-wheel front Eric Oliver also went to smaller wheels during the year with 3·50 × 16 in. Avon tyres with a square section, car-type, tread. These were fitted to car-style alloy rims and he continued with the offset axle fork, hollow spindle front hub and swinging fork rear suspension. One of his major opponents, Cyril Smith,

Ray Amm's 1954 500 TT winner in the Dundrod paddock before Ulster practice and still carrying its TT number on the side of the proboscis

preferred the rigid layout, so his machine had tubes added to lock the rear wheel spindle in one place.

Oliver suffered a crash and a broken sidecar wheel spindle which spoilt his world title chances but Smith took over to snatch the title from the Gilera drivers once more for himself and Norton. Duke kept his 350 crown with four classic wins in a row before a crash at a lesser meeting put him out for the rest of the season. In the 500 cc class Reg Armstrong rode valiantly but found, like Duke, that the Italian fours were becoming too fast, too fine of handling and too reliable to be beaten on sheer cornering and rider ability. Duke went on to ride for Gilera while Guzzi took over the 350 cc class, but Norton still had Oliver riding on three wheels.

During 1952 an interesting experimental engine fitted with a rotary valve was built up at Bracebridge Street. This came about because Laurie Bond, designer of the Bond mini-car and a monocoque two-wheeler, exhibited a rotary valve head on a 500 cc JAP speedway engine. Joe Craig liked the look of it, and took an option.

The engine that was modified was a square works unit with 86 mm stroke and Bond worked on the project at the Norton factory. The valve on the JAP engine was chain driven but on the Manx they used the vertical shaft with a bevel on the end of the valve. This lay across the head so the carburettor stood out on the cam drive side and the exhaust on the drive side. To get the valve in the space the vertical shaft had to be stepped out a little and this was done by a pair of spur gears in a re-vamped lower bevel housing. They were housed to act also as an oil pump and supplied a copious feed of oil to the two large bearings the valve ran in.

Right **Rear engined sidecar built by Ernie Earles for Bill Boddice to drive in 1954**

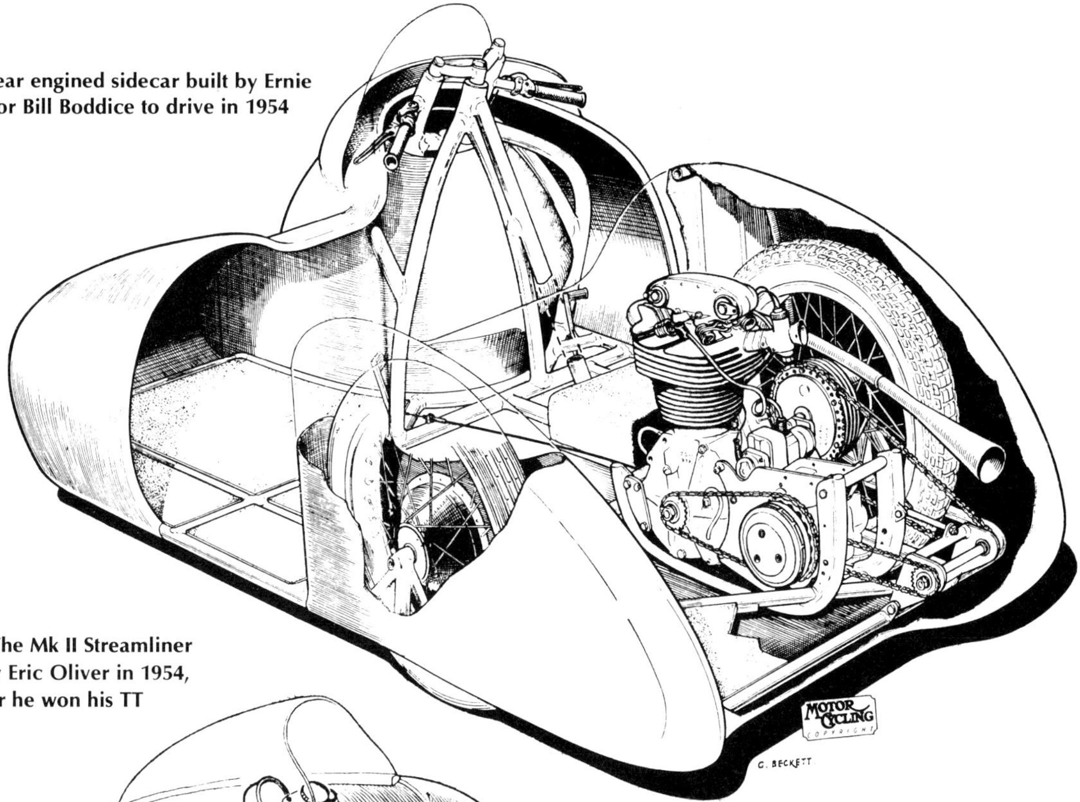

Below **The Mk II Streamliner used by Eric Oliver in 1954, the year he won his TT**

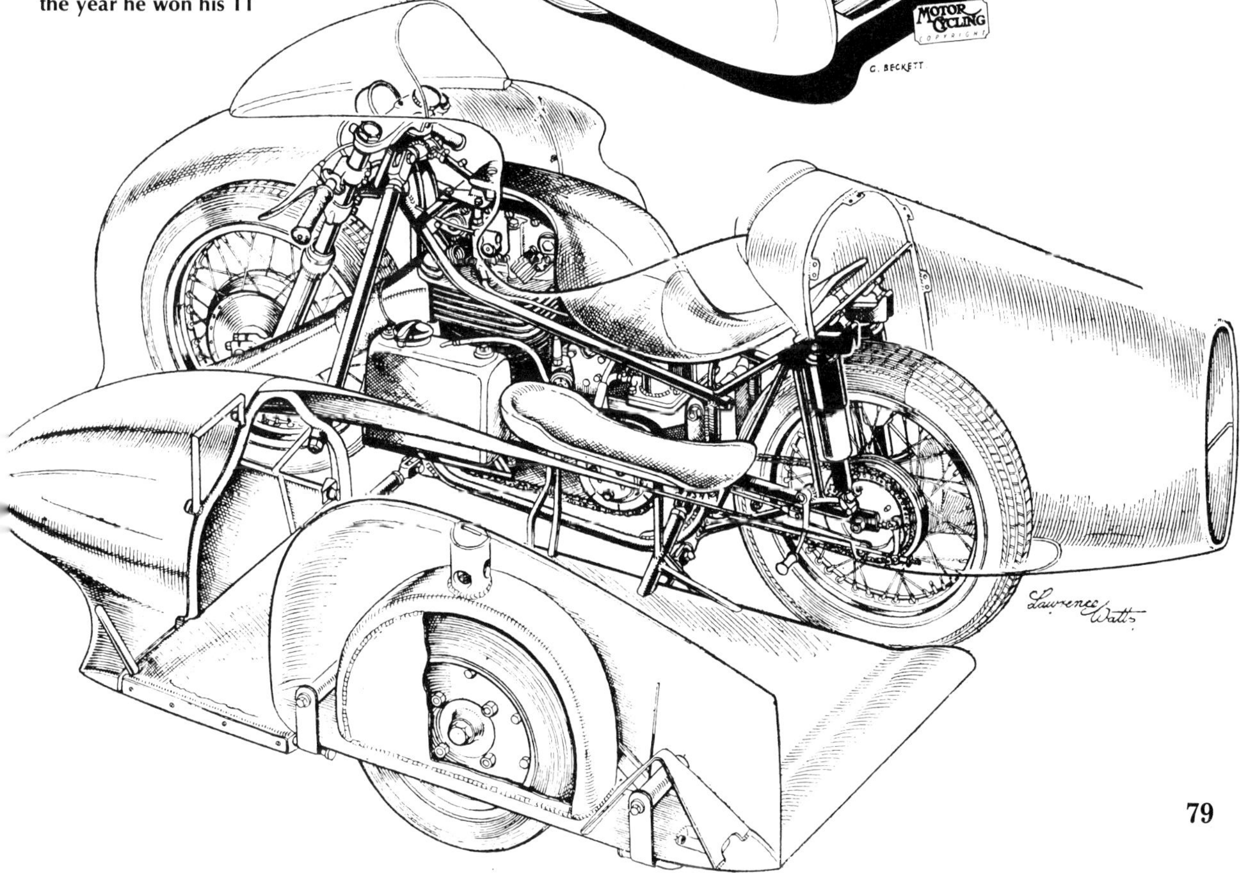

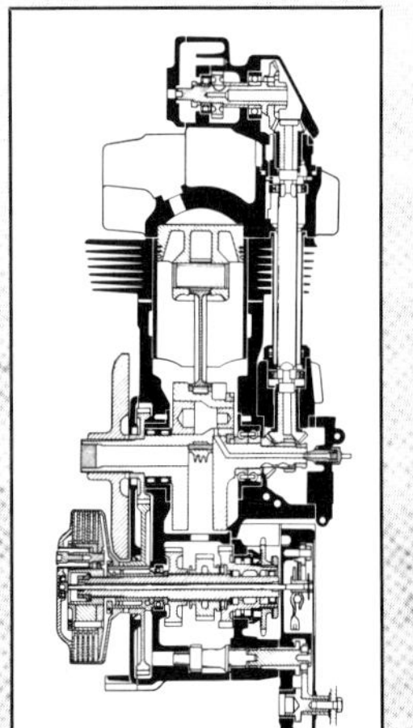

The experimental F model built for 1955 but never raced

Right Cross-section of F engine and gearbox

The beach at Daytona with a Norton heading a group of riders

Below **The 1954 works 350 with two piece crankshaft, special mains, finned bevel housing and outside flywheel**

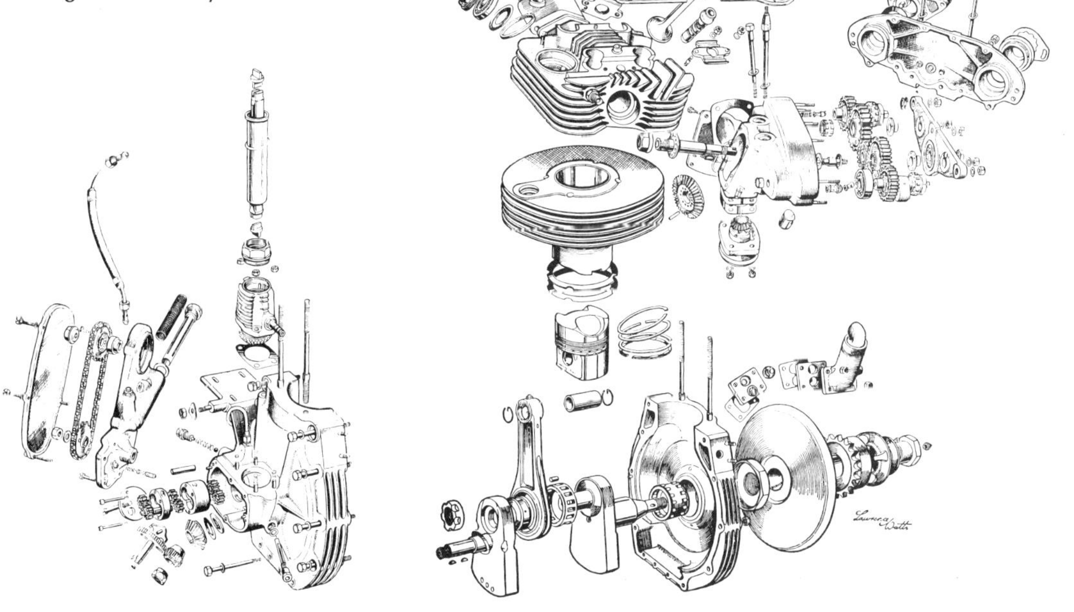

A Manx Norton which has had the Francis Beart treatment both inside and out

It was an interesting experiment and in time the power output went up to 47 bhp or about the same as the normal twin cam unit. However, it experienced many seizures on the way and even when these were overcome it still had problems with over oiling when shut off, which prevented it from picking up cleanly. At the time it had been thought that the engine could have been used in 1954 but before then the factory changed its policy and development stopped.

This was not the first rotary valve Norton for back in 1947 the Brown brothers of Wednesbury converted a new International to such a layout following a design tried out in 1938. They used the normal camshaft drive to the rotor which lay across the cylinder head with the carburettor behind it in its usual position. The exhaust came from the rotor end on the left with a massive curved and well finned adaptor connecting to the pipe and silencer on the left. When tested by

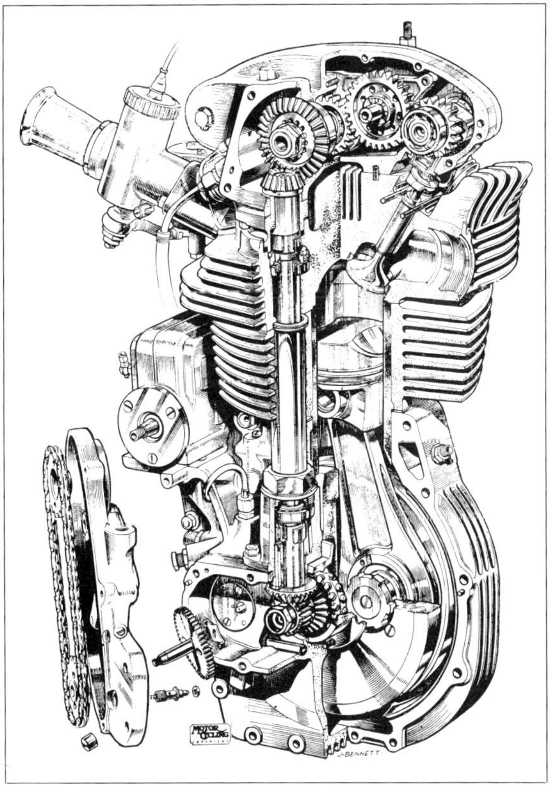

Above **Has to be a works special with AMC box, Inter engine, odd oil tank and bent silencer**

Left **The Manx engine for 1957 with coarse pitch bevels and sodium cooling for both valves**

The Motor Cycle the machine performed well up to the 90 mph region, returned 70 mpg when driven hard, and ran very quietly without the normal Inter clatter.

Even more unusual was a chain driven cammy Inter conversion made just after the war by E. Jackson of York using a 1936 model 30. The conversion was done by welding sprockets to the machined bevels top and bottom which retained the vernier coupling to the camshaft but reversed its direction of rotation. The chain drive was enclosed within an alloy two-part case which attached to the lower bevel housing and sealed against the cam box. It contained two ballraces against which the chain ran and one was movable to set the tension. Coil valve springs were fitted as the casing prohibited the use of hairpins. Mr Jackson also fitted Dowty air forks and a short arm rear suspension system with constant chain tension device to his machine, which was still running in 1960.

1953 brought no changes to the two Manx Nortons but heralded very revised Internationals. Away went the iron engine and garden gate frame to be replaced by an all alloy unit in the featherbed.

The adoption of the swinging fork frame was expected as the touring models had changed from the old plunger and the twin, in sports form, had used the new frame for some time. In fact the new Inters were really the model 88 Dominator twin fitted with the camshaft single engine. The gearbox was changed to the laid down type, the silencer to the pear-shape adopted by several models that year, the saddle for the dualseat with tool tray underneath, and the front brake to the 8 in. size. The new brake had an iron drum rivetted to the hub and continued the Norton tradition of pivoting each shoe on a separate pin.

The electrics were still the faithful Lucas mag-dyno and the lights copied the road range with underslung pilot lamp and Diacon rear. Their control switch and the ammeter were mounted with the speedometer in a panel attached to the top of the fork legs. The dynamo control unit went into the tooltray and access to this was by

Jack Brett with the proboscis plus undershield in 1957

removing the qd seat.

The oil tank size was increased a little and the petrol tank held down by a central strap in the standard featherbed manner. Mudguards were sporting and close fitting and pillion rests fitted as standard.

The two engines retained their traditional dimensions and single camshaft but the head became a light alloy casting with iron valve seats and the barrel an alloy sand casting with an iron

Right **Modified 350 engine used by Mike Hailwood fitted with twin plugs and coil ignition**

Below **Jack Brett on the Slazenger Norton**

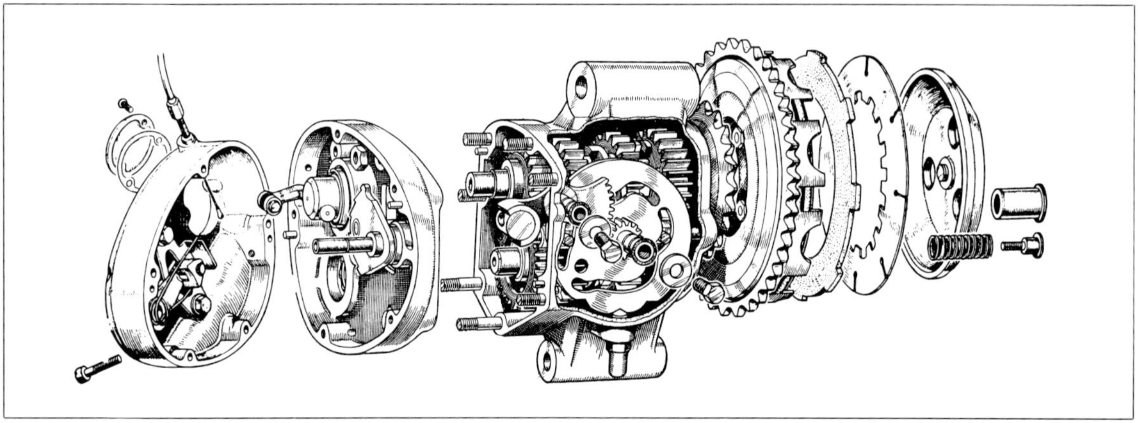

The racing AMC box adopted for the Manx from 1958

liner. Compression ratios remained as they had been at 7·3 and 7·2 for the two sizes. As in the Manx of earlier years the oil was fed through a jet screwed into the cambox instead of through a groove machined in the camshaft sleeve. This allowed the oil to be more easily led to the cam flanks and the point where it was most needed.

TT carburettors were used and the inlet port was at a downdraught angle. No air cleaner was fitted.

A range of extras were offered for the Inters which converted them to what was an old type single ohc Manx but in the later frame. Thus the owner could buy racing footrests, gear pedal with linkage, rear-mounted brake pedal, folding kickstart lever, small mudguards, racing handlebars, a flyscreen and a racing exhaust pipe.

Any ideas of continuing the process to produce a true Manx were not really possible as the Manx crankcase was deeper and stiffer than the Inter's, while the frames also differed in that the rear units of the Manx were vertical while those of the Inter sloped forward a little as on the twin. The real difference was that the frame material for the racing frames was always Reynolds 531 while the road models had to make do with a commercial grade.

The works racing team had an unusual machine to try out in March 1953, the Norton Kneeler which became known as the 'Silver Fish'. This was a very lowly built 350 with extensive front and rear streamlining in which the rider laid in the manner of later day sidecar outfits. It was tried at Montlhèry and over the Mountain section of the TT course ridden by Ray Amm and Ken Kavanagh, but kept under an all-enveloping cover away from prying eyes as far as possible.

The machine made its racing debut in the North-West 200 ridden by Ray Amm where it made a poor start, set a new lap record the second time round, and retired at the end of the third with engine trouble.

The Kneeler had its normal works 350 cc engine in the usual position in a frame based on the featherbed. This was modified in that the top tubes were splayed out to pass either side of the cylinder head and ran to join the lower tubes just above the rear pivot point. An upswept subframe supported the rear units and the fairing. Fuel was carried in pannier tanks on each side of the engine and pumped up to a weir-type carburettor. The pump was driven from the inlet camshaft and the same arrangement used on standard models with weir carburettors to return the overspill to the tank.

The streamlining of the 'fish' was very low and extensive with the front nose fitted with ducts to direct cooling air to the engine. The enclosure

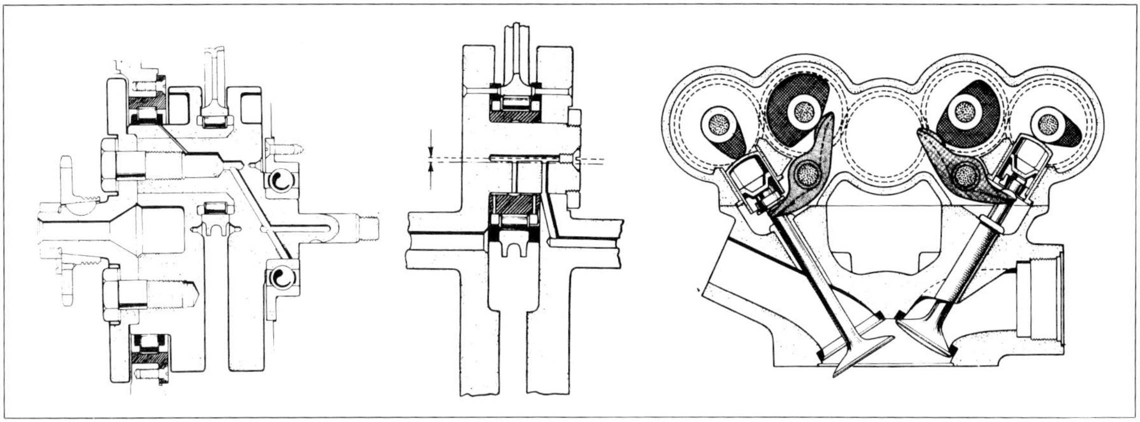

1958 innovations from the left were the big bearing crankcase, the 93 mm bore engine with eccentric crankpin sleeve and the desmodromic cambox

came down to a point just above the wheel spindle and its line was continued along the sides by the tanks which were bridged under the frame by further panelling. At the rear of the tanks were formed kneeling recesses for the rider's lower leg, and at a later stage armrests were added to the tops of the pannier tanks and a header tank also fitted. Padding went above the oil tank and formed a seat for the rider with the backrest attached to the tail fairing. This ran to enclose the rear wheel again down to spindle height and was of a width to allow the rider and his legs and feet to be within its outline.

On its first outing Amm had found it ran round the corners fairly well and was very fast down the straights. Its next outing was in practice for the TT where it recorded some very acceptable times and headed the leader board on one occasion. By then it had gained a second cooling tube on each side of the nose and was referred to as the 'Amm Sandwich'! In the end he used the normal machine for the race, which he won, and no more was seen of the Norton fish in road races.

He also had the task of trying a novel front fork with short trailing link suspension at Floreffe on the 500 cc machine. This design had the spring units mounted behind the fork and attached close up to the link pivot. It also featured a sprung and deeply valanced front mudguard clamped to the fork tube and this must have cost it a few mph from the extra drag. The fork was tried again in TT practice on both sizes of machine, but in the end was not raced.

The works machines used that year had oil cooled exhaust valve guides with a gilled tube clipped to the right downtube and connected to the cambox and the bottom bevel chamber which was finned. Weir carburettors were used with the return pump driven from the inlet camshaft, while the larger model had another dimension change to an even shorter stroke at 88 × 82 mm to allow more revs without raising the piston speed.

On the transmission side an experimental gearbox with more rigid shafts and redesigned clutch was tried, the clutch having more torque capacity but no shock absorber which was moved back to the engine shaft. The primary chain oiler had twin feed tubes to the links and its guard was extended round the clutch to keep oil away from the rear tyre. The rear chain oil was kept in the left subframe support tube.

To keep the oil temperature down, the tank was painted matt black and, to keep the rider's knees in, the rear of the fuel tank was modified. A return was made to 19 in. wheels at both ends with a 3·00 in. front tyre and 3·50 in. rear.

The golden days were over, however, as although Amm did the double at the TT and Norton won the two major classes in the Ulster, world titles eluded the solos as did other classic victories. Not on three wheels though, not while Eric Oliver continued with his single for he won four of the five classics to take his fourth title.

The outfit he used began the season in the same form as the previous year but with a lowered steering head and the fuel carried in a tank alongside the rear wheel. From there it was lifted to the carburettor by an electric pump powered by a battery carried in the sidecar nose. A shroud replaced the normal tank and overall the racing height was down five inches.

This was just an interim step for in July came a fully streamlined outfit with semi-prone riding position on the lines of the solo kneeler. The nose of the fairing was a double shell with the main air flow taken away from the flat front number plate and the air over that ducted to each side. The tail was very similar to that of the faired solo and the sidecar wheel was also enclosed. The tank mounting on the chair floor and the pump were retained and the machine and chair faired together on top and underneath.

It was a good year for the sidecar Norton as Cyril Smith was 2nd to Oliver four times and won in Ulster, while the race leader boards were dominated by the marque with only a couple of 3rds going to the BMW.

1954 was the year that the Manx took its final form with revised engine dimensions and welded-on subframe. The prototype was first seen at the Manx GP the previous year and, when the range was announced, it confirmed the details.

Bore and stroke became nearly square at 76 × 76·85 mm to give 349 cc and 86 × 85·8 mm to give 498 cc. The connecting rods were shortened and valve diameters increased to take advantage of the larger bore. The finning copied that of the works engines with that of the cylinder encompassing the vertical cam drive tube, the top housing becoming integral with the head with

Below **A 500 cc Manx ridden in the 1959 Temple 100 by Bob McCracken**

Right **Geoff Duke at the start of the 1959 Ulster GP with his 500 on which he finished third**

TWORK
NCHES
26
12
12

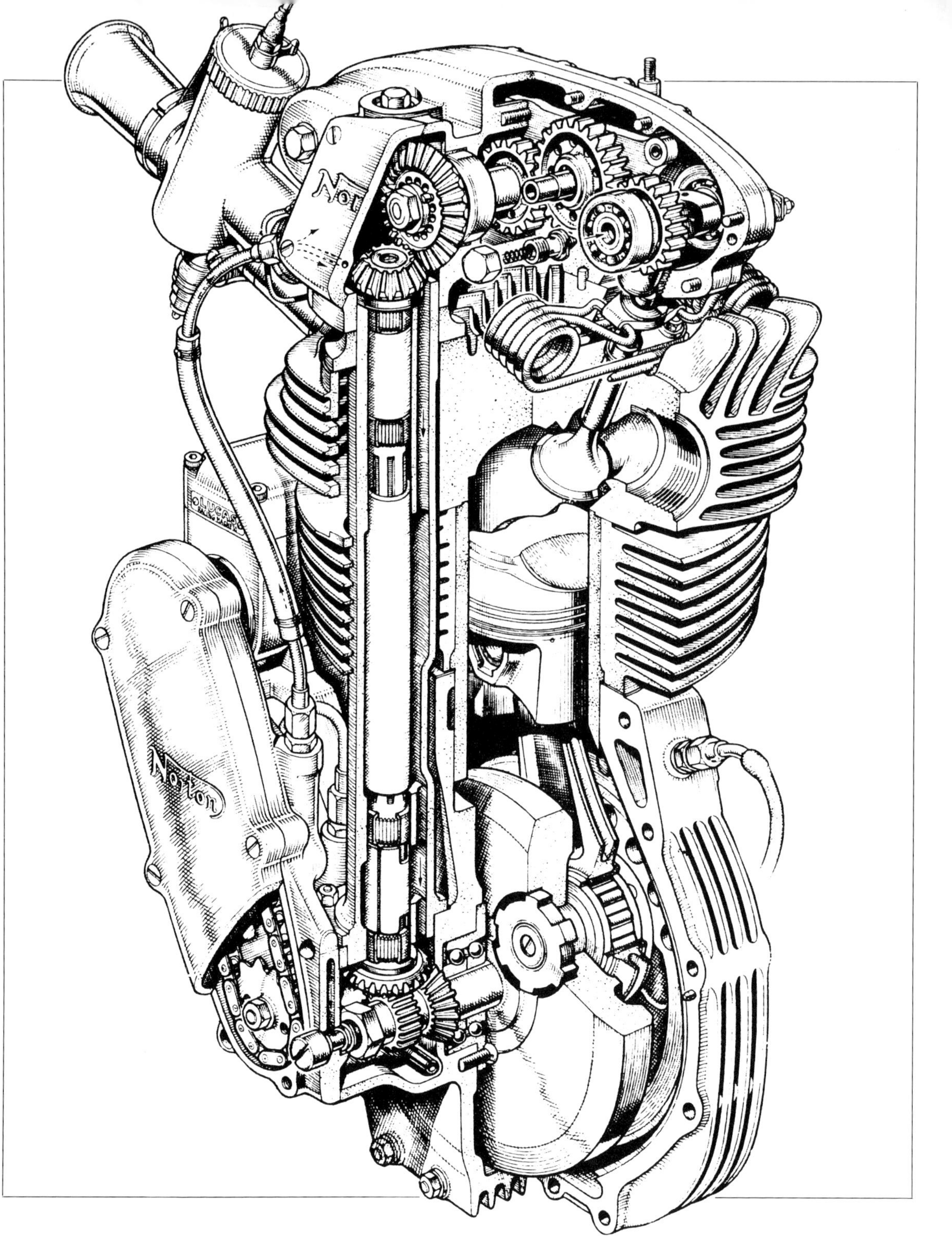
LUCAS
Norton

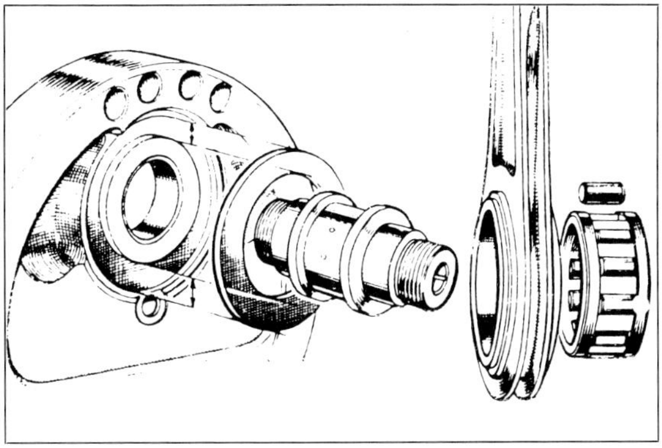

Eccentric thrust washer adopted in 1960 to take rod side load and unable to rotate

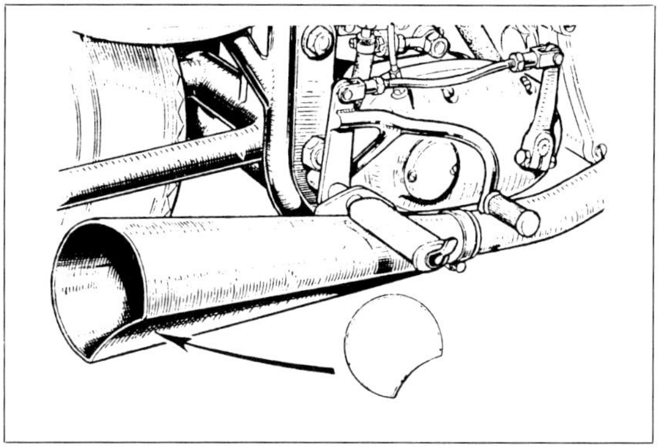

The fluted flat of the 1960 megaphone used to combat vibration on the 500

fins round it, and the adoption of the diagonal fins on the top of the head above the combustion chamber and between the valve guides. The cambox no longer sat on pillars but directly on two machined surfaces on the head and a pair of dowels. The valve angle was altered a little and the cambox modified to suit the new bore sizes. The inlet tract angle steepened.

On the cycle side the main change was the welding of the subframe to the main loops, but of equal importance was a twin leading-shoe front brake. This retained the earlier 8 in. diameter and the conical hub but copied the design of the works backplate. This carried twin cams linked externally and two fixed pivot posts for the shoes in a magnesium alloy casting. Posts and cams were linked by a steady plate as before and an air scoop was fitted. A change was made to the oil tank shape to copy the 1953 works type but it retained the normal finish.

With all this work going into the Manx it was hardly surprising that the Inters remained unchanged as they had been extensively altered for the previous year.

The works machines were also well altered for 1954 with even shorter strokes at 78 × 73 mm and 90 × 78·4 mm for the two capacities, the 500 being known as the 90 bore from then on. With the revised dimensions came shorter rods, a stiffer crankshaft and an outside flywheel. The crankshaft comprised just two pieces, for the mainshafts were integral with the webs and the crankpin with the drive side. The timing side was pressed into place after the big-end bearing inner race had been shrunk on and the caged roller and rod assembled. A castellated nut secured the crankpin and was part machined away after assembly to clear the main bearing housing. Both mains were in flanged supports, the timing side with a double row ball bearing and held by six screws. The drive side had the flanged sleeve pulled up by a large external nut and the double row caged roller bearing ran direct in the sleeve and on a hardened inner sleeve shrunk onto the mainshaft.

Out-board of this went the plain flywheel and next to that the engine sprocket which had a vane-type shock absorber with spheroidal rubber elements built into it. Older riders recalled that Francis Beart had built a 350 cc outside flywheel engine back in 1949 for the Manx GP. He had put the flywheel out-board of the sprocket.

The crankcase and its timing side assembly of camshaft and magneto drives was little changed but a new crankcase breather with twin flap valves was mounted on the front of the case

Left **The final form of the Carroll engine with needle races for the splined vertical drive**

halves just below the top face. The external connecting pipe contained an oil trap and drain.

The top end was also to a now very familiar pattern and the valve included angle was 64 degrees. With compression ratios over 10:1 the pistons were of the slipper type with flat tops around the valve cut-aways to give a squish effect with the head, which had a small dome, surrounded by a flat ledge between the valves to match the piston. A single plug went on the left a little nearer the inlet than the exhaust, and for one event twin plugs were tried but gave no advantage in lap times, although bench power was up a trifle.

The head fins between the cambox supports ran straight across and the inlet tract angle was 20 degrees. Otherwise it was the story as before with a small AC fuel pump driven by the inlet cam nut which was machined with an inclined face to act as a cam and push the pump plunger.

The engine drove a new five-speed Burman gearbox which was narrower than normal, special to Norton and fitted with a direct pedal with no linkage. It carried a five-plate clutch and the primary chain lubricant went into the top left frame tube.

On the chassis side the lower rail was altered to clear the flywheel, the subframe modified, the rear damper shrouds discarded and, on the 350 only, the weight reduced by going down a gauge on the main loop tube thickness. Braking received attention with the rear operated by a cross-shaft through the fork pivot, pedal on one end and cable arm the other. At the front the hub grew an extractor fan to assist airflow. Shorter exhaust pipes were used to match the increase in engine speed allowed by the stroke reduction.

A new form of fairing was tried with a nose that extended out to the front of the front tyre. It became known as the 'proboscis' and was used with side panniers which concealed pannier tanks on occasion, their contents moved to the top tank by the AC pump. The weir float chamber was no longer scavenged.

All these changes brought Ray Amm a controversial Senior TT and a couple of wins with the 350 but no world titles. The firm was also out of luck on three wheels although Oliver continued with his Streamliner in mark two form. In this it had a small diameter sidecar wheel and the fuel tank built into the sidecar mudguard. Up to then FIM regulations had prevented the use of a small chair wheel by insisting that it was as the others. Oliver retained the 16 in. machine wheels and much of the earlier fairing but this lost the double skin nose.

He failed to retain his title due to a crash but won three of the races and finished 5th in the Swiss with an arm that had re-broken due to a spin. Perhaps inspired in part by the very extensive fairing on Oliver's outfit, Ernie Earles went one further and built a well-streamlined machine with a rear mounted Norton engine for Bill Boddice to drive. The engine and gearbox went into the cut off lower half of a featherbed frame and drove a small countershaft mounted at the rear of the chassis frame from where a chain drive ran forward to the rear wheel.

1955 brought a change in the official Norton racing policy for they felt that it had become too remote from normal machines and too expensive. Thus they would use standard Manx machines without fairings but incorporating features for developing for the following season's models. Only selected events, mainly held on pure road circuits would be contested.

Thus the works men rode 1955 Manx Nortons which were as the previous year but with various details on trial. The engines had minor cam and port changes, weir float chambers and, most noticeable, rotating magnet magnetos. Chassis changes were all minor but the forks had external springs, the rear fork pivoted on phosphor bronze bushes and worked against Woodhead Monroe units with exposed springs, the handlebar controls were welded in place and had spring-locked adjustors for the two main levers, a Renolds primary chain oiler with Amal control

Fron Purslow in the 1960 Cookstown 100 complete with mop-up felt on the frame

jets was fitted and a strut added to the gear lever. The front mudguard was smaller and without stays, the rear one stiffer, the seat lower and softer, the footrests tubular and rubber covered, the exhaust pipe and megaphone welded into one, and a magnetic rev-counter fitted.

For the private owner the Manx continued in its 1954 form, while for the road the Inter was given the welded subframe in line with the sports twin and full width hubs similar but not the same as used by the other road models.

What the real 1955 works machines would have been was not revealed until early 1960 and even then they caused quite a stir. The machines were still singles but with the cylinder horizontal, as on the Guzzi, and with an integral five-speed gearbox. The oil was carried in the main tube of the spine frame and the machines were known as the 'F' type, this standing for 'flat'.

The engine followed works Norton practice in most of its details with the familiar bevels and shaft drive to the twin camshafts in the cambox. While this was to the normal pattern, the cylinder head was not and its finning was more extensive and the exhaust pipe was held by a flange secured by bolts rather than the usual Norton threaded ring. Both engine sizes were built using the 1954 works dimensions and they retained the outside flywheel on the left. The crankshaft followed the same lines of construction but with the big-end rollers running directly on the crankpin integral with the drive side.

The cylinder retained its normal radial fins but the bottom bevel housing was finned axially into the airflow. The housing carried the oil pump as usual but in place of the magneto was a contact breaker assembly for coil ignition with manual advance. The battery to power this went into a tray just above the flywheel. The oiling system was modified to suit the flat layout and the oil cooler linked to the exhaust valve was retained.

Primary transmission was by spur gears to a novel five-speed gearbox with two layshafts and cross-over drive. The clutch ran dry in the air but the gears were enclosed by a light cover. The gearbox was very narrow, as each shaft only had to carry three gears, and was short as the layshafts were positioned above and below the mainshaft. The final drive was on the right.

The frame was based on a single large diameter tube with the headstock welded to one end and the engine support plates to the other. It also doubled as the oil tank. The engine unit was held between these plates and supported by twin head steady tubes which bolted into place. Further pairs of tubes formed the seat and rear unit support and an alloy plate and a casting carried the swinging fork. The casting also enclosed the gearchange and clutch mechanism so was part of the engine unit assembly.

At the front went telescopic forks and both wheels had twin leading-shoe brakes, the front with the 1954 type extractor fan and the rear with remote sprocket.

The fuel tank would have been of a size to suit the race and the pump might have been needed on occasion as the carburettor was above the engine at a steep angle. It was an Amal GP specially modified to work at this angle with an adaptor at its base to carry the main jet near the vertical.

On test the F gave as much power at the rear wheel as the 1954 works motor had at the crankshaft so, with its low build, it could have had the right potential when one recalls that Guzzi riders continued to win the 350 cc title with their flat single up to 1957.

At the end of the year Joe Craig resigned from Norton as he intended to remarry, settle abroad and take a well-earned rest. His contribution to the Manx had been enormous and it was tragic that early in 1957 he was killed in a car accident on icy roads. It was thought he suffered a heart attack while driving, but it was later found to be a broken rib, caused by the crash, which went into the heart. At 59 his loss to the world of racing was immense.

The 1956 Manx Nortons fully bore out the firm's policy for they copied the machines run by the factory team the previous year. In addition to the items already listed, other points were a new flanged double row ballrace on the timing side with the flange secured to the crankcase by screws, full ring clutch discs, an extended primary chainguard, a chain guide or buffer for the final drive, a smaller flyscreen, no air lever and the primary chain lubricant in the top left frame tube. Inside the engine the exhaust valve, which had had a hollow stem for sodium cooling since 1951, gained a hollow head and more sodium.

The works machines were seen at the TT and incorporated further detail changes with slight increases in choke sizes on both models and lower height achieved by the use of shorter fork stanchions. The sub-frame was modified to delete the rear loop behind the seat and the rev-counter was recessed into a rubber mounting in the top of the number plate fairing which was itself rubber mounted to the forks. The front brake drum was ribbed for strength and cooling.

The Inter was no longer listed after 1955 but continued to be hand-built to order on a very limited basis indeed. It always retained the laid-down Norton gearbox but in other respects was given the road model changes for ease of

production. Thus for 1957 the fuel tank was just painted with separate chrome plated side panels with a plastic beading round their edges, each panel carrying both the badge and the kneegrip. The seat was modified and a new tubular silencer fitted with offset inlet and no tailpipe.

Because so few Inters were made they began to use Manx parts where it suited the firm, so some were built using the racing frame in 531

Left **Engine room of the 1961 Manx showing new oil tank fixing, revised megaphone clip and stop to prevent it sliding off**

Below **A Norton prepared by Steve Lancefield and fitted with the 1962 dual front brake**

tubing with the extra lugs added on and the rear fork with phosphor bronze bushes. Engines also began to appear with a mixed pedigree, so few, if any, of the last models were really built to the Inter parts schedule.

The Manx was modified with a redesigned connecting rod with a sleeve in the big-end and altered bearing and crankpin, a new cam for the 350, slightly bigger carburettors, sodium cooled inlet valves, and the camshaft drive bevels changed to coarser pitch teeth.

There was to be no works team for 1957 but the proboscis fairing with pannier tanks and undershield made an appearance with an ex-works machine underneath and three 90 bore 500s ran in the Senior TT. These were a development of the 1954 engine and had all their flywheel inside the crankcase as the external one tended to fracture the drive side mainshaft. A longer rod was fitted to ensure the piston skirt cleared the wheels, and the combustion chamber was modified.

There was a rumour of an engine with desmodromic valve gear at the TT but, although such an engine was built, it only ran on the test bed. It used the intermediate gears in the cambox to carry the closing cams which moved rockers to reach under collars on the valve stems. At first the desmo engine had standard cams but was later run in 1958 with much fiercer ones. The result was a fractured vertical drive shaft which led to a redesign in that area which appeared on the Manx in 1959.

The major change to the Manx for 1958 was the gearbox which adopted many of the features of the AMC box used on the road models since 1956. As on the road one, the racing box retained the essential features of the old Norton unit with the pawl and ratchet positive stop mechanism of the AMC one. The racing box had the clutch set in closer as no chaincase was needed, ball bearings at both ends of the layshaft, an improved camplate profile and an outer cover shaped to allow the exhaust pipe to tuck in against it. The clutch mechanism was a ball and cam lever device rather different to the road models but, like them, required that the clutch pressure plate had an adjustor at its centre. Rather a step back as the big mushroom went and with it the automatic square lift. The clutch cable entered vertically and the gear lever connected to the pedal with an external linkage.

1958 saw the end of the Inter—'not with a bang but a whimper'—a rather sad end to such a long and illustrious history spanning a quarter of a century with a straight linear descent from 1932. They had become too special, too expensive and too difficult to service in an age when the trend was to bigger vertical twins.

At first the works policy was as in 1957 but they then gained sponsorship from Slazenger and several special engines were built. One was a Junior big bore with 80 × 69·5 mm dimensions and was to be used with a five-speed gearbox by Geoff Duke in a special frame with leading link forks. Coil valve springs from America were fitted with a view to enclosing the valve gear and these springs were to become very favoured over the years by private owners.

In the 500 cc class Doug Hele had two engines and the first had 93 × 73·5 mm dimensions. The stroke reduction was achieved in a classically simple way by making the big-end inner race eccentric and pressing it on to the pin to give the required offset.

The second 500 retained the more usual engine dimension but could be drastically reduced in stroke if this was required. It was known as the big bearing crankcase, or BBC unit, and was easily recognised by a circular cover fitted inboard of the engine sprocket and enclosing an external flywheel. In fact the driveside main was a gas turbine roller bearing with a bore of no less than 5·5 in. This enabled the crankpin to lie inside its bore so the construction was that the timing side flywheel with integral

Right **Catalogue page for the 1962 Manx models in the last days of Bracebridge Street**

Manx 30/40

The most successful standard production racing machine in the world.

ENGINE. Both engines are of basically similar design employing bevel driven double overhead camshafts; light alloy heads; "Alfin" barrels; electron crankcases; high tensile steel connecting rods and crankshafts; and forged light alloy pistons. The compression ratio of 11 to 1 is suitable for International racing fuels. "Lucas" rotating magnet magnetos and "Amal" G.P. type carburettors are fitted.

TRANSMISSION. Close ratio four speed gearbox together with a three-plate clutch employing bonded inserts; primary chain oil is carried in the frame tube.

FRAME. Double loop, bronze welded "Reynolds" 531 tube Brackets are fitted for the attachment of a fairing.

SUSPENSION. "Roadholder" telescopic front forks employing hydraulic damping; swinging fork rear controlled by "Girling" suspension units.

WHEELS. Light alloy serrated rims, electron hubs, brake plates and shoes. The new front brake is a "twin" type employing two leading shoes on both sides. The surface area is 71½ sq. in. and large air scoops are fitted.

TANKS. Light alloy, rubber mounted, with quick action filler caps.

FINISH. Black and silver enamel with chromium plating.

Models	Bore and Stroke (mm.)	Cylinder Capacity	Compression Ratio	Gear Ratio	Weight	Petrol Tank Capacity	Oil Tank Working Capacity	Tyre Size	Brake Dimensions
30M	86 × 85·6	499 c.c.	11 to 1	Top 4·23 2nd 5·63 3rd 4·65 Bot. 7·52	313	5 Gal.	7½ pts.	F.3·00 × 19 R.3·50 × 19	Front: Double 7″ × 1⅜″ wide
40M	76 × 76·7	348 c.c.	11 to 1	Top 5·12 2nd 6·81 3rd 5·64 Bot. 9·11	307	5 Gal.	7½ pts.	F.3·00 × 19 R.3·50 × 19	Rear: 7″ dia. × 1¾″ wide

mainshaft had a substantial crankpin also integral which was bored and tapped. The drive side flywheel was a disc with a large diameter and very short mainshaft which fitted the roller bearing bore and an even shorter crankpin on which the big-end ran. This was bored so it could be fitted onto the timing crankpin which ran right through the whole of the driving wheel to give a very rigid structure. It was held together by a bolt screwed into the crankpin and this, with other bolts screwed into the drive side flywheel, clamped an outside flywheel to the inside one.

This engine was tried in a sidecar outfit for ease of installation and ran well in the TT but retired for reasons not associated with its BBC.

1959 saw the works desmo 500 and a very short stroke 350 with dimensions of 86 × 60 mm tried in TT practice but not raced. The Manx had the vertical drive shaft altered to that developed for the desmo, with a pair of needle races supporting top and bottom bevels located in a single tube bolted to the bottom bevel chamber and sealed into the cambox with an O-ring. The drive between the bevels was by a tube with internal spline at each end which mated with the gear shafts. Thus away went the bottom bevel housing and the Oldham couplings, while the drive shaft tube diameter was much larger than before. The camshaft bearings were altered and the valves made solid without the sodium cooling, the exhaust in a Nimonic steel. Several private owners fitted a second plug either under the cambox or behind its drive. On the cycle side, the clutch had its inserts bonded to the driven plates and one less plate fitted.

For 1960 there were quite a few changes for the production Manx, many internal, while on the outside was a new seat pan and front number plate cowl in fibreglass, a perspex flyscreen, a 9000 rpm rev-counter, an oil tank held by a strap, and a flute in the megaphone flat to stop it vibrating and splitting. Also on the outside the gear pedal linkage was altered, the rear wheel sprocket dowels strengthened and internally serrated wheel rims fitted that allowed security bolts to be dispensed with.

On the inside was a wider big-end eye for the connecting rod and recessed bars for the light alloy big-end cage. To accommodate the added width the side thrust washers were made thinner and eccentric so that each was trapped in its recess and unable to turn. This allowed the earlier fixing screw to be deleted along with its tapped hole.

The piston crown was raised along with the compression ratios from 10·6 and 10·07 to 11:1.

Above **Tickle Norton minus tank but showing off the high level exhaust to good effect**

Left **Two T5 models, one in Tickle frame, the other more standard**

The piston was lighter, fitted with a chrome plated top ring, and non-rotating circlips which keyed into a recess in the piston side. In the crankcase the flange screws for the timing side main were changed to cap screws and a magnetic drain plug added. The tappets in the cambox altered from being case hardened to the use of Stellite tips, and the top housing spigot of the cam drive shaft was made deeper. The float chamber had a damped weir flow arrangement.

On the transmission the clutch centre was reinforced with a brazed-in washer and the locking washer for the retaining nut itself retained by a circlip.

The works took some interesting new machines and engines to the Isle of Man for the 1960 TT, while to help the private owners along, Harold Daniell had a five-speed cluster for the AMC box and in later years an Austrian, Schafleitner, was to produce a six-speed conversion, as did Rod Quaife in Kent. The most striking works model had a normal 350 cc engine with normal ground clearance but much lowered machine height as the rider's chin rested only an inch above the cambox.

The machine was immediately christened the 'Lowboy' and, while the general frame was normal featherbed, the steering head was not as it would have obscured the rider's view. Thus a cycle-type design was used with the fork legs carried in a massive bottom crown machined from solid steel and turning in a 3 in. diameter bearing. A small header tank went just in front of the abbreviated seat and fed the float chamber attached to it. It was supplied by pump from the main tank which went into the space under the seat and between engine and rear wheel. The oil tank went below it behind the gearbox and this space was enlarged by welding the fork gussets to the back of the frame tubes and thus moving the pivot back. A shorter fork kept the wheelbase within normal limits.

The exhaust pipe ran under the engine in a recess in the sump and the megaphone pointed at the rear tyre. The fairing was so low at the front that the machine number had to be painted on each side of the nose and still extended on to the screen. An 18 in. rear wheel was used and the footrests moved well back. With the low bars this pointed the rider's toes down, hence the megaphone move as they clashed, but it was a tiring machine over the bumps or when braking. It was practised by Eddie Crooks but not raced.

Norton also had a 350 cc version of the desmo engine with an ultra short stroke with dimensions of 86 × 60 mm. It had an outside flywheel and made a brief appearance in practice only.

Also news that year was the return of Eric Oliver in the sidecar race with a straightforward kneeler outfit with leading link forks, factory engine and full fairing. Unfortunately he crashed in practice, and the race favourite was out.

The Lowboy made a further appearance late that year with a modified riding position and a chest support for the rider to lean on. In 1961 at the TT none of the experimental machines appeared but a dual 7 in. front brake was used and the works effort went into the Domiracer. Later in the year the twin frame was tried with a special 350 engine fitted to it. This had a shorter rod to get it in the frame, was mounted lower and closer to the gearbox than usual, and had coil ignition with the points outboard of the oil pump.

The 1961 Manx had quite a few detail changes mainly in the engine. The piston was lighter and fitted with a single Dykes ring, the barrel shorter, the cams and valve seats altered, the ports larger, the inlet tract longer, the inlet valve lighter with thinner head and smaller stem, the tappets returned to the case-hardened type and the cam tracks hard-chromed. The tappet guides were longer and the valve spring cup and collar made integral to prevent the valve turning. The carburettor became a GP2 and larger with a supporting bracket and the megaphone was fitted with a large clip at its end and a small stop so it would not fall off even if loose. On the cycle parts the oil tank was new and pivoted down from rubber mountings to be held against a pad by twin rubber bands. The rear fork pivot was made tubular to reduce weight.

Manx production came to a halt at the end of 1961 but some machines were built from spares in 1962 and a few the following year, after the works had moved from their hallowed home in Bracebridge Street to the AMC works in Plumstead following an amalgamation.

The 1962 models still had changes, the most obvious being a dual 7 in. diameter front brake, each side having twin leading-shoes. Internally

The 4-valve engine with close set camshafts based on Cosworth technology and built by Syd Mularney

the oil pump capacity was increased and a larger feed sent into the crankshaft, the carburettor size on the 350 increased while the inlet tract spacer had a taper bore on the 500, that model having the petrol tank cut-away at the rear to give more clearance for the bellmouth. The cylinder barrels were improved aesthetically from 1961 by machining both ends to shorten them instead of taking all the material from the top and thus thinning the top fin.

That was the end of the Manx Norton but not the end of the racing singles for during the 1960s riders and tuners continued to use them and develop them. In this period Ray Petty tried a special Joe Craig cylinder head from the mid-1950s which was itself inspired by an Australian who raced a twin carburettor Inter. That engine had its mixing chambers set at a wide angle and with a very steep downdraught, but the 'Joe' head from a 90 mm bore 500 had them nearly horizontal but angled round to the point where they were close to feeding into each other instead of the port. Petty soon returned to a single carb head but reversed it on a 350 to give a rear facing exhaust port and high level pipe.

Spares became very hard to come by in this period and then in 1966 AMC sold all their racing hardware to Colin Seeley. He was more interested in the AJS and Matchless racing singles so the Manx side lagged and was then sold in 1969 to John Tickle who had a thriving business making accessories for himself and the trade.

Tickle set up to make spares and also built complete machines, as the 350 cc Manx T3 and 500 cc T5. These retained the late type Norton engine, the AMC gearbox but had them housed in a Tickle-designed frame very similar to the featherbed. A very noticeable feature was the exhaust system which ran back on the level of the port alongside and just under the seat.

The next and currently final step came in 1970 when Syd Mularney first built a four valve head for a 500 Norton and it gave 58·5 bhp at 7000 rpm straight away. The combustion chamber layout owed much to the racing Cosworth and the engine required a good deal of detail development work before it was fully sorted out, but late in 1978 after some years of limited use it all came together and the machine became very successful in classic bike racing.

It is a fitting end to an engine that had so much success for so many years in the world classics.

4 Overhead and side

Alongside the glamorous overhead camshaft singles, Norton built a range of solid, no-nonsense road models with a pedigree that stretched back to that first prototype of 1907. Changes there were from that early belt drive machine but always the Norton signature was written clearly on the design.

And so it was in 1931 when the new, and for all intents final, singles made their appearance. The range built during the 1920s had progressed but little in appearance and to the end retained the long and tall vintage frame, the magneto out front of the engine, albeit under a large and polished cover, and relied on a somewhat vague lubricating system. So, while Carroll worked on the ohc unit, Edgar Franks drew up the other new models under his guidance.

The essence of the changes was that dry sump lubrication was adopted and the magneto moved to the more fashionable, stylish and safer position behind the cylinder as was already used by the ES2.

The models in the range were the side valve 16H and Big 4, and ohv models 18, 19, 20, 22 and ES2. The 16H, 18, 20, 22 and ES2 retained the 79 × 100 mm dimensions, the Big 4 its 82 × 120 mm, and the 19 its 79 × 120 mm. The model 20 was a twin port 18 and the 22 a similar ES2 for which the owner paid a small additional price for the changed head and extra exhaust pipe and silencer.

All the engines were of the same design and

The 1931 16H with dry sump, rear magneto and odd exhaust pipe line

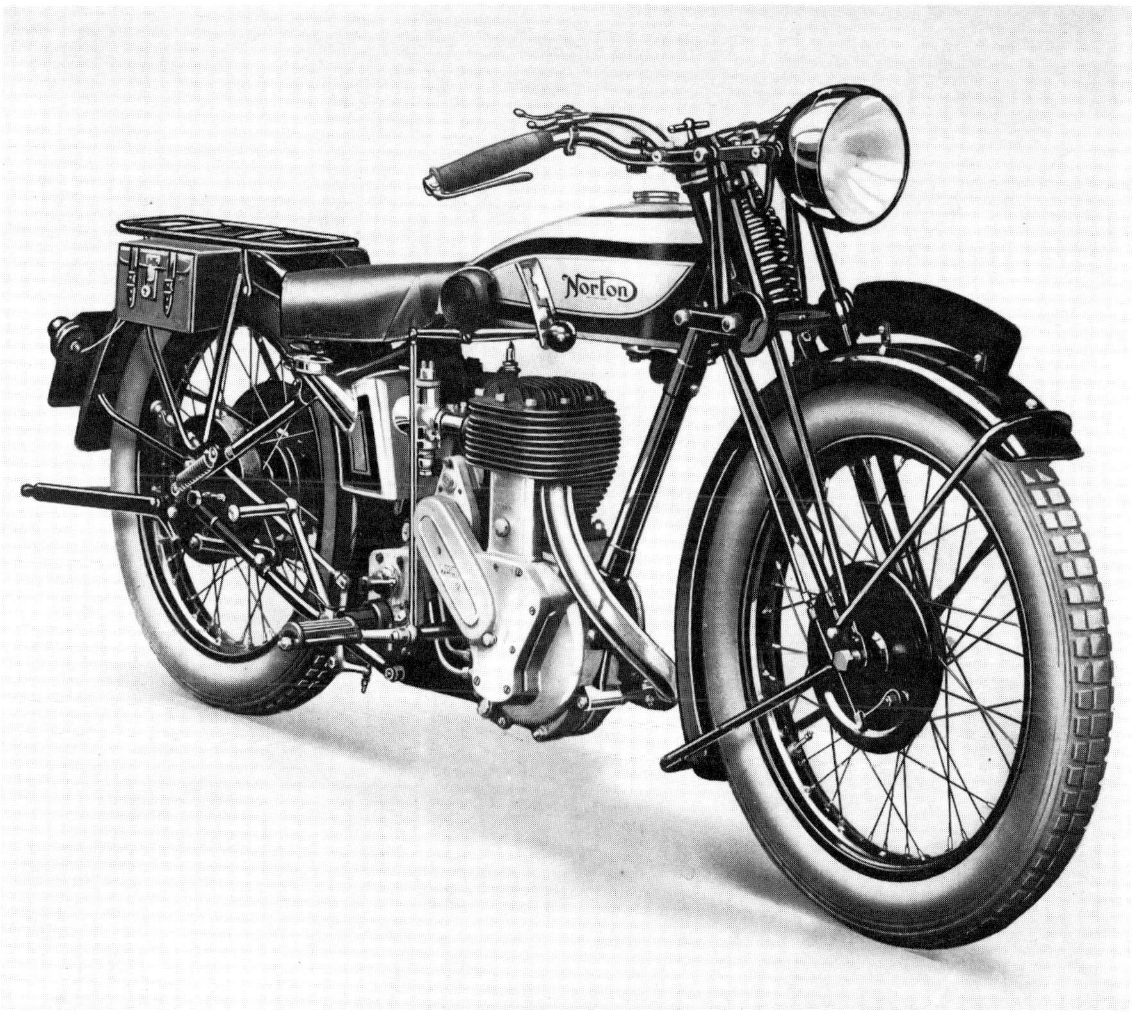

many parts were the same then and for a good few years. The engine was based on a vertically split crankcase which had a small sump in its base at the rear to collect the oil. A cast-in knife edge helped to remove it from the flywheels as these turned. The crankshaft was built up with the mainshafts pressed into the large and heavy flywheels and the crankpin, with its integral shoulders for the double row uncaged roller big-end, pulled up by nuts locked by small screws with a choice of two holes. The crankpin was hardened on the roller track and the con-rod carried a hardened ring pressed into the big-end and a bush in the small-end. The piston was conventional with a scraper and two compression rings and was fixed to the rod by a fully floating gudgeon pin retained by circlips.

The piston ran in an iron barrel held down on studs fitted into the top of the crankcase and also spigoted down into it. The spigot was machined locally fore and aft to clear the rod. The cylinder head was also cast iron and held down by nine studs on the side valve models and four on the ohv ones. The side valve engine used a gasket and retained the old type blanking plug over the piston to enable its movement to be measured. No gasket was used on the ohv units and these had the sparking plug on the left at an angle, while that on the side valve sat over the inlet valve.

The crankshaft turned in a pair of good sized mains and drove the valve gear and oil pump from the right side. There were minor diffferences in the machining of the right crankcase half, depending on the form of valve operation, but the bulk of the timing chest remained the same. A pinion was keyed to the crankshaft and mated with inlet and exhaust camshaft gears in one with their cams and shafts. The shafts ran in bronze shouldered bushes fitted into the crankcase and the timing cover. The cams moved finger cam followers which also pivoted in bushes set in the crankcase and timing cover and it was only above these that there were differences.

On the side valve models a pair of tappets ran in guides screwed into the top of the crankcase and these transferred the cam lift to the valves and had adjustors built into their top ends. The valves worked in pressed-in guides with grease nipples for lubrication and were returned by single springs retained by collars and cotters. A simple lever with return spring pivoted in a bracket mounted on the crankcase top between the tappet guides to act as a valve lifter and the valve gear was shielded from the owner's gaze by a steel cover held by a single screw.

The overhead valve models retained the cam followers which acted directly on the pushrods which had been extended down a year before when the intermediate tappets had been deleted. At the same time the pushrod return springs had gone, along with the 'spectacle' frame cast into the cylinder side that they abutted against on the open rod models and the enclosing covers of the ES2 and JE from which they gained their model numbers.

The pushrods worked on rocker arms in one piece with their shafts which oscillated in grease lubricated bushes in the rocker box. At each shaft left end was secured an external arm on a key and a taper and this arm moved the valve via the screwed adjustor at its outer end. Twin valve springs were fitted. The rocker box was held down by four bolts and had a side cover on the right attached by six screws. This cover enclosed the rocker arm and pushrod top and was designed to clamp seals to the top of the pushrod tubes. These tubes attached to the crankcase with unions and nuts.

For lubrication Carroll at first intended to use the camshaft engine type gear pump but Franks argued successfully that in mass production it would be much better to have a unit that just bolted to a flat surface, not into a close tolerance hole. So the famous Norton gear pump came into being, bolted to a face in the timing chest beneath the crankshaft from which it was driven by a threaded worm wheel that also secured the

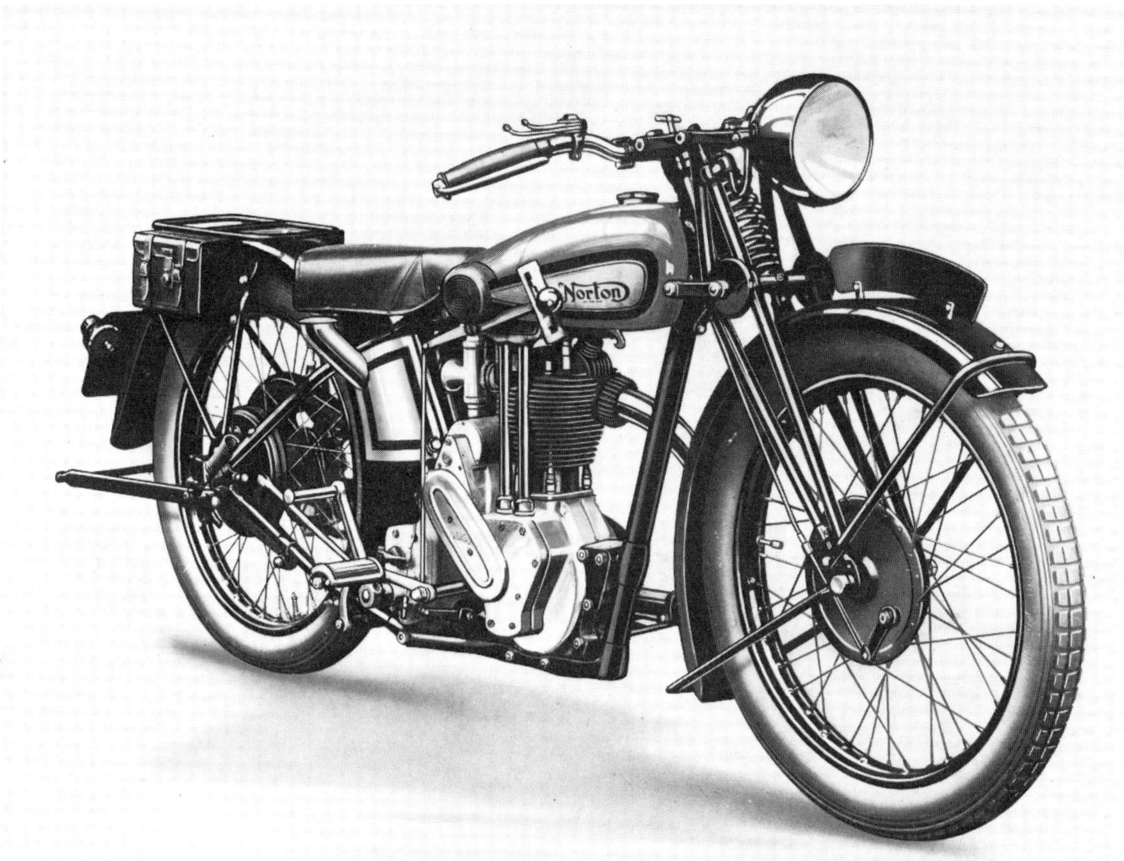

Above **1931 ES2 with cradle frame and rear brake on right side**

Below **1931 Big 4 with cast footboards and fishtail silencer**

timing pinion. The pump had two gear pairs and was to be used for the singles and, in a different arrangement, for the later twins.

The pumps were made by Abingdon and were very tolerant of the fit of the worm drive despite the apparent tolerance build up between the two parts. The only fault in the design was a tendency for the oil to drain into the sump if the machine was left for a while. Up to 1930 a tap had been fitted in the feed line to the engine but the penalty for forgetting to turn it on was somewhat final. Norton maintained that oil did not leak past the pump gears, but owners found it did and cursed the lack of a ball valve.

From the pump the oil went via a nipple into the timing chest and drillings to the points needing it. A relief valve controlled the pressure and allowed any excess into the timing chest which carried a plunger tell-tale in its top surface to assure the owner that the oil system was working. The main feed was by spring loaded jet into the end of the crankshaft from where it went via drillings to the big-end. This item was not drilled and the oil was fed into the side of the bearing. A take-off from the feed supplied oil to a drilling in the rear of the cylinder wall and an early screw adjustor was soon altered to be permanently set.

Oil from the engine drained into the sump in the crankcase from where it was collected by the scavenge pump and returned to the oil tank. A single filter was fitted to the feed line in the tank.

Ignition was by Lucas magneto with a third brush dynamo strapped to its back to make a mag-dyno. The combined unit was mounted on a platform behind the cylinder and driven by chain from the inlet camshaft. Enclosure of the

Above right **1935 model 20, the 490 cc 18 with twin-port head and two silencers**

Below **Classic Norton side valve, the 1932 Big 4 with silencer on right. Only details changed in its remaining 22 years**

chain was achieved by extending the timing cover up to the magneto and fitting an outer lid secured by screws.

Carburettors were Amal in one of two sizes with separate float chamber and the side valve models had exhaust pipes which curled round the front of the crankcase to run along the left side of the machine to the silencer. All models had a left silencer and the twin port models the extra one on the right.

The engine had a cam lobe shock absorber on the left mainshaft and drove to a Sturmey Archer clutch via a single strand chain under a pressed steel cover. The clutch also had a shock absorber which was built into its centre and was a dry running, multi-plate type with Ferodo inserts and six springs. It drove a three-speed Sturmey Archer LS gearbox with hand change for which there were alternative ratios available. Owners also had the option of a four-speed, hand change gearbox.

Two types of frames were used with all except the ES2 engine going into a diamond frame that

Small brother to the 20, the 348 cc twin-port model 55 in 1934

Above **1936 model 50 with fatter pushrod tubes which improved appearance**

Top **Mid 1930s trials specification with narrow mudguards and raised exhaust but otherwise little changed**

was a shorter version of the two rear stay type. The ES2 used the three-stay cradle frame from the CS1 and also had the oil tank from that, so differed from the remainder of the range which had rather square tanks without the curved lines reserved for the more sporting models.

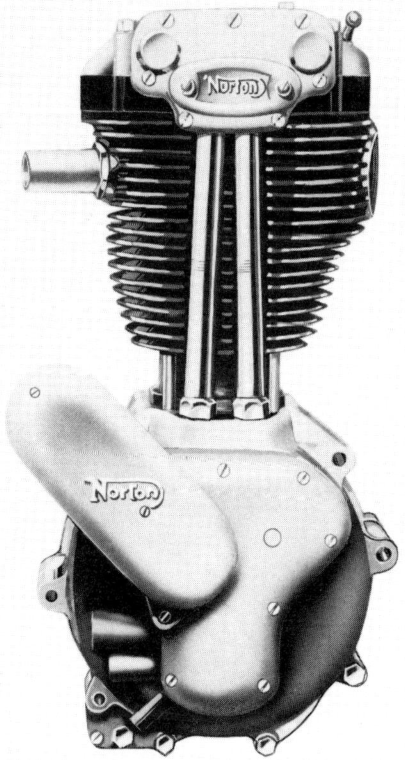

Above **A 1937 bigger banger, the 597 cc model 19 with common frame**

Left **The 1938 engine with enclosed valve gear and modified cams with flat followers**

Forks were Norton and based on the Webb but without the useful knurled thrust washers that made adjustment so much easier. The far-too-easily qd Norton wheels were tried but found wanting, so Enfield hubs went at the rear, with right side rear brake on the 16H, and Horton or Webb at the front.

The remainder of the cycle parts were standard for the era, with saddle tank, heavy steel mudguards with the tail of the rear one detachable with the number plate to assist wheel removal, a saddle, pillion pad if required, battery on the left, headlamp, rear stand and toolbox. The side valve models still retained their vintage style rear carrier and leather toolboxes hung from it, but these went the next year and the toolbox became steel and went between the chainstays on the right behind the oil tank. It was also given a hinge at the top of the lid with the catch at the bottom to drive owners to distraction as they fumbled for tools in the dark while

The 1938 16H complete with very effective but most unsightly silencer used for that year only

trying to hold the lid open and up. It remained so for some years but was reversed by 1936.

As well as the steel toolbox, the side valve models also had their exhaust pipes moved onto the right side for 1932 which, with the fishtail silencer, made for a much better appearance. The major change for the year concerned the wheels which adopted the three stud Norton design used until well into the 1950s and full width hubs. The design gave interchangeable wheels and the brakes became the famous Norton type with 7 in. drums and $1\frac{1}{4}$ in. wide alloy shoes, each on its own pivot post. The separate posts gave the shoes very ample support and, if bridged at their outer ends, the braking performance could be improved still further from a good standard.

The rear chain was the exception to the industry and remained so to the end. Its narrow width gave all the strength needed and it seemed that the malleable iron of the rear drum and integral sprocket was kinder to the hardened chain rollers than the normal steel others used. Certainly it gave no trouble.

1933 brought a change in engine dimensions for the model 19 from the 79 × 120 mm of yore to 82 × 113 mm which gave 597 cc, although it was always quoted as 596. The bore was thus the same as the Big 4 side valve model which retained the 120 mm stroke. Two new models appeared in a smaller size as the single port 50 and twin port 55. Both had dimensions of 71 × 88 mm to give 348 cc and were otherwise the same as the models 18 and 20.

The transmission remained Sturmey Archer but all the ohv models were fitted with four-speed gearboxes as standard, although still with hand change. The positive stop was available as an option. The side valves continued with three speeds with the option of four. The wheel hub centres were enclosed to hide the three fixing nuts and an end cap went over their ends, but this was not a very successful move as out of sight nuts could be slack and not noticed until the hub was damaged. Also the end plate collected rain water and fed it to the wheel bearings to their alarm and short life. By 1936 the hubs were naked once more and stayed that way. The silencer fishtail changed its shape to fantail and during the year the model 22 was dropped.

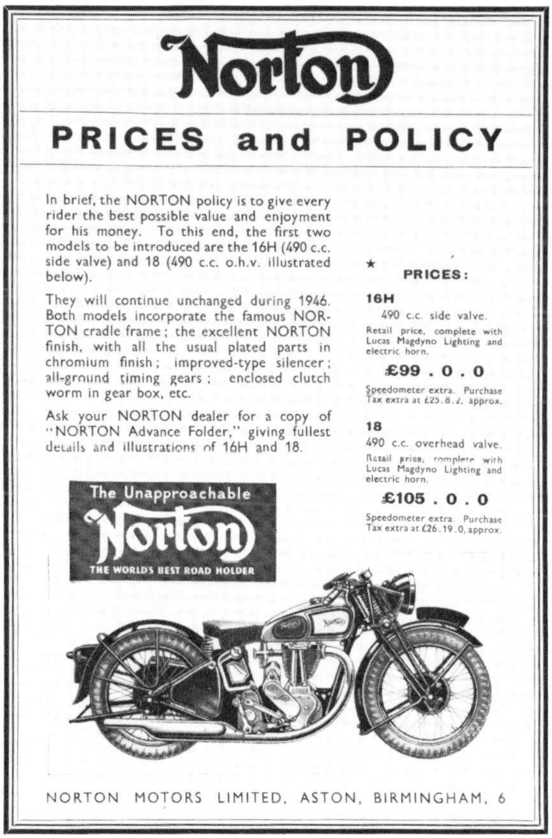

Above **The final prewar form with open diamond frame for this 1939 model 50**

Left **September 1945 advertisement setting out the Norton range for that year and the next**

1934 brought some engine changes with a second main bearing on the drive side and a revision to the timing gears so they ran as a train with the crankshaft driving the exhaust and that in turn driving the inlet. This change from meshing both cam gears with the crankshaft pinion was done in the interests of reducing timing gear noise but made little, if any, difference. What it did do was to reverse the direction of rotation of the inlet cam and hence the magneto. Thus for many years the ohv and side valve Norton engines were almost unique in fitting a clockwise drive magneto. On the ohv models small oil pipes were added from the rocker box to the valve guides. Improved tubular silencers were also adopted, these having the inlet lower than the exhaust and, with minor changes, remained in use into the 1950s, except for 1938.

The machines all lost the engine shock absorber that year and were fitted with the Norton clutch and oil bath primary chaincase. The gearboxes were as for the previous year but the positive stop mechanism was made neater and housed under a smaller cover. On the cycle side the frames were fitted with the forks with check springs and the knurled thrust washers.

The next year, 1935, saw the final major transmission change with the adoption of the four-speed, foot change, Norton gearbox. It also saw the exhaust pipes all on the right side where they were to stay. In addition to the standard models there was a competition option available for most. This used a high ground clearance diamond frame with the narrow fork from the camshaft models and the appropriate non-qd front wheel. To this basic change was added a high level exhaust, folding kickstarter, and narrow chrome plated mudguards.

1936 brought an increase in the oil circulation rate and, due to this, increased diameter pushrod tubes which looked much better than the previous rather spindly ones. On the 500 cc ohv models the top half fixing was changed to four long studs with sleeve nuts, as had been used on the 350 from 1933 and the 600 before then. In 1937 the ohv engines were given 14 mm sparking plugs, although the side valves continued with their 18 mm, and the electrics became two brush and compensated voltage control, the box for this going onto the right upper chainstay. The ES2 continued as a cradle frame version of the model 18.

There were modifications to the engines for 1938 to fully enclose the valve gear. On the side valve models this was done by casting a chamber onto the side of the cylinder for the valves and

1946 16H fitted with low compression engine, hand gearchange, right rear brake pedal and special hand-start lever to suit a disabled ex-serviceman

springs to live in. This was set away from the cylinder so that an air passage remained between the two and an alloy lid acted as a cover. Lubrication of the valves was by a fifth holding down stud that went between them. It was drilled from below along part of its length and then crossdrilled with holes aimed at the valve stems. This let the oil mist from the timing chest through to do its job.

The overhead valve models had a new cylinder head with wells for the valves. On top of each went an alloy packing piece about an inch deep and on top of that went a rocker box which extended over both wells and sideways to the right. On that side went the two pushrod tubes which were angled in at the top but still located by nuts on to unions in the crankcase and by a side cover plate at the top. As the valve adjustors had been moved to the tops of the pushrods a further access plate held by two screws was provided to enable them to be reached. The rockers continued to work in bushes and

The early postwar 16H still with girder forks but in a cradle frame and with chrome dome on chainguard

became three part with each arm held on a keyed taper by a nut. There were no drains to the valve wells in the head, which were just scraped out when decoking.

Down in the timing case the cam gears were made wider, the cam shape altered and flat cam followers fitted to improve the quietness of the engines. The oil tell-tale was no longer fitted which removed one prevalent oil leak.

The oddest feature of the year was the silencer, for the tubular unit was replaced by a hefty rectangular box with twin outlets, one above the other. The works nicknamed it 'the cow's udder', and this was as good a description as any for something with a basic Brooklands can shape but with radiused curves not the usual sharp track outline. A contemporary road test of an ES2 in *Motor Cycling* was not too highly impressed with the new silencer and said that the note was 'just a shade too healthy' but not

The 1947 ES2 with Roadholders, plunger frame and revised gearbox end cover

'raucous' and so all that was required was 'to refrain from too-violent acceleration'. The same machine recorded 82 mph with 71 in third and 53 in second, which was about par for the 490 cc ohv single from 1930 onwards. Fuel was used at 72 mpg and the test concluded that the ES2 was 'real value for money'.

Although the machine was really a tourer the lights were listed as an extra along with the horn, pillion equipment and speedometer. Also available was a tank top instrument panel, which was fitted, this being a popular prewar method of moving the instruments as far from the line of sight as possible. The trials specification was available for an extra £5.

For 1939 sanity returned and the tubular silencer was back on all models. The carburettors were fitted with an easy start device which was a small lever clamped to the throttle stop screw. The idea was that the half turn it allowed would raise the slide the right amount for cold starting so that the rider did not have to judge the throttle opening. It was not used much.

Two options were made available for the ES2, the first the plunger frame as used by the

Above right **1948 16H with revised valve gear**

Right **Side view of 1948 16H showing the smaller timing chest and new form of valve enclosure**

HOV 242
Norton

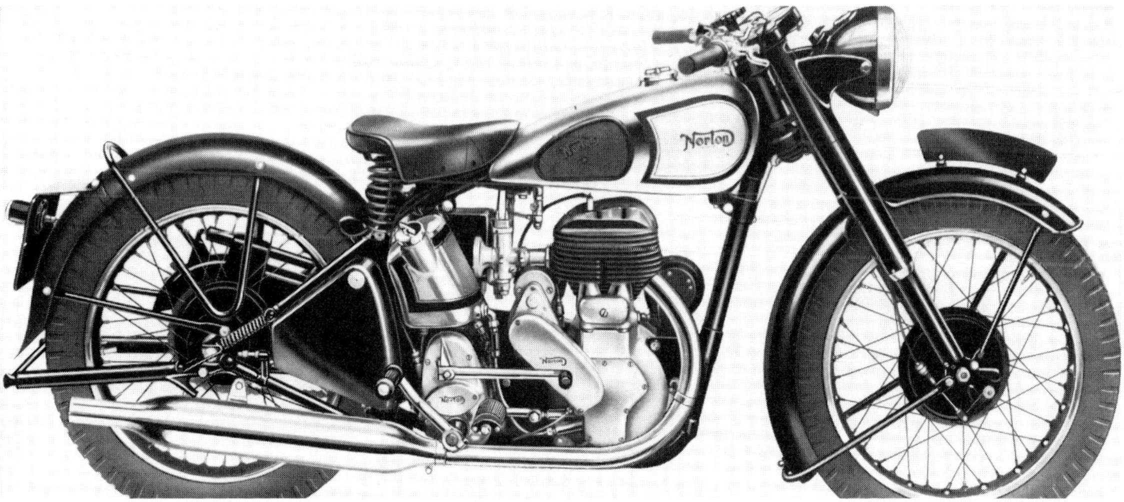
Norton

camshaft models and the second the International type of petrol tank. With the plunger frame came the non-qd rear wheel to match and with the tank came a much improved appearance.

1940 was to have seen the introduction of revised engine dimensions for the 16H, 18, 20 and ES2 to 82 × 94 mm and 496 cc. The prototypes ran smoother and the use of the bigger head, effectively that of the 600, encouraged an increase in the size of both valves and carburettors. The side valve models would have had alloy heads and on all models the gearbox end cover revised to enclose the clutch arm and move the gear pedal to a position where it could be operated without lifting a foot from the rest. It was intended that the ES2, 30 and 40 would continue with their cradle frame, plus the option of telescopic forks as an extra, while the other machines used a cradle frame with twin tubes running forward under the engine and bolting to the down tube.

A further change intended for 1940 was the introduction of 350 and 500 cc trials models built as such rather than as modified standard road machines. They were intended to have all the expected features for competition such as increased ground clearance, short wheelbase, wide gears and suitable cycle parts. Postwar the idea was to become the 500T.

Before that could happen Norton had to turn their attention to wartime production of the side valve models and little else. During the conflict the telescopic fork design was altered to add hydraulic damping and initial trials carried out on war department 16H models among others.

Once the war was over production was the essential issue with a government clamouring for exports and a population at home and abroad desperate for transport. Thus any ideas of changing engine dimensions had to go by the board, along with any other alterations.

The simple answer for Norton was to continue building the 16H and add the model 18 which was so similar in engine details and to use the same cycle parts for both. They did take the opportunity to make some changes so the engines were to the 1939 pattern and the frames full cradle type. Thus the valve gear was enclosed unlike the military 16H which used the 1937 type engine, and the frame was that used prewar for the ES2 and camshaft models. To cope with pool petrol an aluminium head gasket was added on the 18 which lowered the compression ratio a touch and eliminated the need to grind the head joint.

The two machines were fitted with the

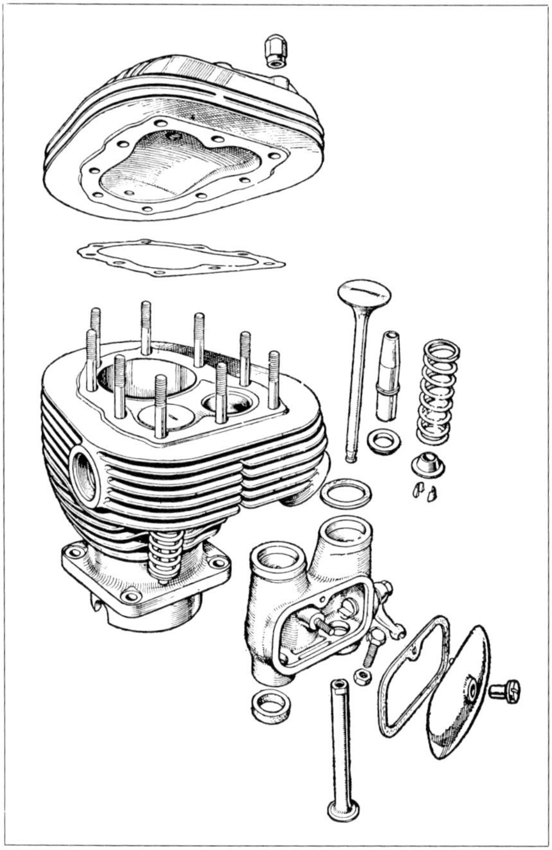

New side valve top half for 1948 with chimney pot style enclosure casting

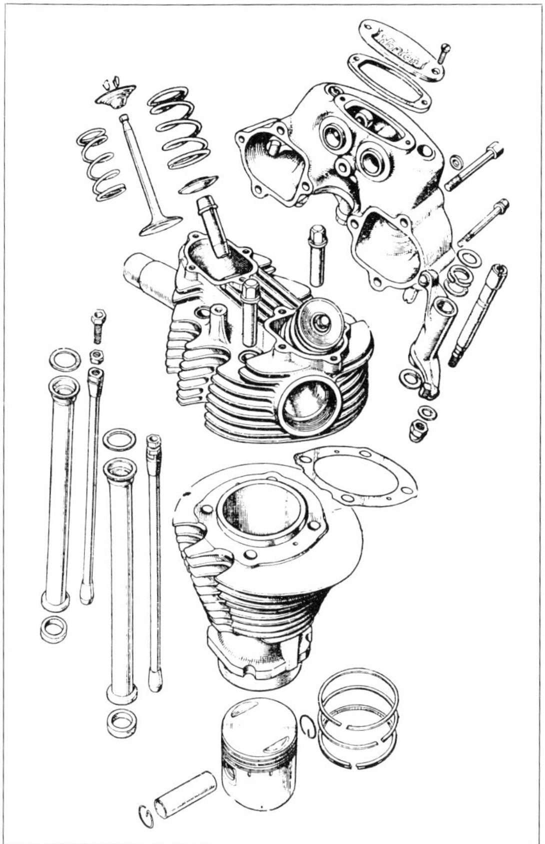

Above **Norton advert from 1951 showing an outfit on patrol with the RAC**

Left **New ohv top half for 1948 with one piece rocker box and fixed spindles**

Below **Rare plunger Big 4, 1950 model restored so has a few non-original parts but not many**

Above **The swinging fork model ES2 from 1953 with pear silencers**

Left **Drive side of the 1953 ES2 which is fitted with the massive Norton prop-stand and has no engine number**

enclosed clutch worm end cover to the gearbox with its pressed steel access cap, a larger toolbox, a revised oil tank and a modified silencer. The machines continued with girder forks, but with the headlamp shell held by stays not the prewar prongs, no rear suspension and in the traditional finish.

1946 saw no changes to the two models except in price, but for the next year the range expanded and was brought up to date with telescopic front forks. These were the famous Roadholders with oil damping and went on to the whole range which had the ES2 and Big 4 re-introduced. The first was as the 18 but with plunger rear suspension supplied as standard, while the second was as the 16H and 18 but with the 634 cc side valve engine. The four models were thus all very similar and had the speedometer mounted in the fork top bridge while the ammeter and light switch remained in a small panel in the rear of the headlamp shell. Some ohv engines were fitted with cams with quietening ramps which were so extensive that compression could vanish when the engine was cold. Such units had the letter 'Q' after their number.

1948 brought a redesign to all the engines despite which the Norton line and style carried on just as always. Also the stroke of the Big 4 was reduced to 113 mm and the capacity to 597 cc. The most noticeable aspect of the changes concerned the valve gear but internally the flywheels reduced their diameter and increased their width which allowed the piston skirt to lengthen. The old cam followers went so the timing case became smaller and in their place went tappets with flat feet. The back face of each cam gear was dished to accommodate these and allow them to rotate. This did not prove to be totally satisfactory so the tappets gained a flat on one side of the foot to prevent rotation. Around the same time the number of teeth on the magneto sprockets changed from 15 to 16 and later rebuilds from assorted spares which failed to check for this could result in the phenomenon of an engine that fired once and then not again for quite a long time.

In the side valve engines the tappets carried adjustors at the top which bore directly on to the valves. These and their springs were enclosed by a one-piece aluminium chamber with the appearance of two chimney pots rising from it. This chamber was clamped between crankcase

Above **The 250 cc single built in 1953 with high camshaft ohv engine and now at Beaulieu**

Below **Last of the side valve line, the form of the 16H and Big 4 in 1954 showing their linear descent from the first belt drive single**

Above **The Norton stand at the 1955 Earls Court Show with the new 1956 models on view**

Below **The AMC gearbox adopted by the road models in 1956 and directly descended from the prewar Norton unit**

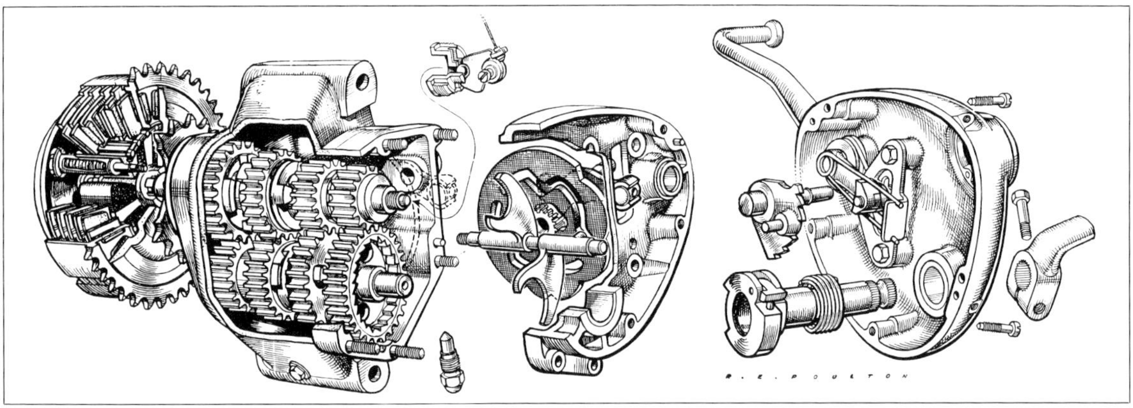

Final rigid model, the 597 cc 19R of 1955 with fat rear tyre to aid comfort lost when dualseat replaced saddle

and cylinder barrel with sealing washers at each end and stood well clear of the cylinder wall. An access plate held by a single nut allowed the tappets to be adjusted and the casting also carried the valve lifter arm. A further feature of the revised side valve engines was a light alloy cylinder head with a 14 mm plug screwed into a bronze sealing cast into the aluminium.

The overhead valve engine had the same changes to flywheels, piston and valve gear but with the top of the tappets rounded off for shorter,fatter and lighter tubular pushrods. Their tubes were clamped between crankcase and a new one-piece rocker box with a seal at each end. In the box the rockers were shorter, stiffer and worked on fixed spindles lubricated with a feed taken from the main oil return pipe just before it reached the tank. This fed a banjo on the left side of the box from where internal drillways connected to the rockers and in turn to the ball ends that rested in the adjustable cups at the top of the pushrods. The valve wells in the head were given oil drains back to the crankcase via drillings in the cylinder. Finally, the ohv models were fitted with the in-line tubular silencer that the sv and Inters had used from 1947.

Changes were minimal for 1949 as the works concentrated on the new twin so all the singles received were the brake shoe pivot tie plate and the rear light moved to the centre of the number plate. Internally, alloy pushrods went into the ohv engines.

The Motor Cycle had an ES2 on test early in 1949 and it reflected the machine and the range well, little changed over the years. Top speed came out at 78 mph and it could be cruised at an effortless 70 up hill and down dale without any trouble. Less enticing was a short note of a bad vibration period between 60 and 70 in top which in the journalism of the time must have 'shaken the tester's fillings out!'

The gearbox and clutch worked very well indeed, while the brakes were outstandingly good. Roadholding was typical Norton—firm and sure with the accent on staying on line even if this was at the expense of some comfort. Overall, and on pool petrol, a machine able to carry its rider at relaxed high speed with fast averages coming up easily without either party having to work at it.

Much the same applied to a test they carried out a little earlier on a 16H. A little slower at 73 mph with vibration from 55 to 65 so cruising was down outside this band. All the other attributes were there so again it was a machine for deceptively easy averages.

For 1950 the gearbox was changed to the laid-down type and the top mounting altered to suit extended engine plates. A common junction was used to attach the oil feed and return pipes to the engine and the voltage regulator tucked away into the toolbox. On the ES2 a prop stand was added and the rear chainguard extended.

New and larger petrol tanks with 3.5 gallon capacity and new oil tanks which retained their size were specified for 1951. Prop stands were fitted to all the models by then and the front brake was modified with a die-cast light alloy backplate.

1952 brought no changes and a *Motor Cycling* road test of a Big 4 with double adult chair summed up its performance as 'lusty'. Not so much in top speed, which was a bare 60 mph,

Smallest single, the model 50 of 1958 replete with brakes and weight of the 600

The ES2 in 1957 with full width hubs and styled battery box

The biggest 1957 single, the 19S with matching style to oil tank and common chassis

but more in its traditional ability to plough on at more or less the same 45 mph regardless of the number of passengers or size of the load.

There was more alteration for the ES2 in 1953 when it went to a swinging fork frame based on the old type brazed lug construction with single top and down tubes, not the featherbed. At the same time a smaller exhaust pipe and pear-shaped silencer were fitted and the gearing altered with a rear hub with 43 teeth. The hub was identical with that used for the rigid models which were all fitted with dualseats and new tail lights. These went onto the ES2 also which had the doubtful pleasure of the inept underslung pilot light at the front. The toolbox was tucked into the corner of the sub-frame on the left and its lid secured by a one-armed wing nut. This was a trick learnt in the war as, if slack, the weight of the arm stopped it from undoing further.

A road test by *The Motor Cycle* showed the ES2 to be up one mph to 79 but otherwise much as before with rather hard suspension and solid brakes. It retained some vibration but also the ability to cruise along quickly enough to cover the ground well.

In February 1953 came the announcement that Norton were to merge with the AMC group

Traditional 600 role pulling a chair along, done by the Big 4 in the early days

at Plumstead in London but would continue as a separate concern, while the rivalry in competition and business would carry on. Gilbert Smith continued as managing director at Norton and no-one could have cared more about the firm than him; his whole working life had been bound up with Bracebridge Street and its fortunes.

There was just one change in 1954 but it applied to the whole range which was fitted with an 8 in. front brake. What would have been news if it had come to production was a new and smaller single of 250 cc. The prototype of this was built in 1953 with a high camshaft overhead valve engine whose cylinder inclined forward a little and had working dimensions of 66 × 72·6 mm. The crankcase carried a 1954 number 57850 plus the marking Mk V, and was vertically split in conventional manner.

Onto it went an iron barrel held on four short studs and to this was bolted the combined head and rocker box in iron. The camshaft was installed high in the timing case on the right and worked pushrods, each in its own tube, which were splayed out from case to head. The rockers lay across the head and so aligned with the valves set into wells with alloy lids to them. Ignition was by coil with a small points housing set behind the cylinder and driven from the timing side.

The gearbox was bolted to the rear of the engine and both box and timing covers were styled into one. The primary chaincase was the normal Norton pressed steel assembly and a rear chaincase was fitted.

The frame had swinging fork rear suspension and a single top tube but duplex down, cradle and seat ones. Telescopic forks went at the front and the machine parked on a centre stand, there being no side one. The engine was dry sump

Featherbed ES2 in 1961 so slimline in addition to alternator and coil ignition

lubricated with the oil tank in the usual place on the right matched by a battery on the left under a cover. Behind these items went twin toolboxes, one each side. The carburettor was an Amal 276 and the exhaust went on the right to a pear-shaped silencer.

Hubs were not full width and brakes were single leading shoe, while both rims carried 3·00 × 19 in. tyres. Mudguards were stock so the rear retained its usual hinged section at the back to assist wheel removal. A dualseat was fitted and the tank held on two horizontal studs at the front and a single central fixing bolt with rubber insulation at all points.

Controls were conventional and the head-lamp shell had a separate panel with the combined lighting and ignition switch on the left and an ammeter on the right, while the speedometer was carried in the fork bridge as usual. The finish was all silver-grey with chrome plated wheel rims with silver centres lined red.

Late in 1954 another unusual Norton was reported in the press, but in this case not a factory job but one built by a private owner, one Freeman Sanders of Penzance. He was an authority on diesel engines and decided he would like to try the principle on a motorcycle engine.

To do this he took an ex-WD 16H bottom half and added an ohv barrel and his own cast iron head with vertical valves. The injector pump was CAV and fitted in place of the magneto and the injector itself, also a CAV item, went into the left side of the head in the normal plug position. The dynamo drive was re-worked to come from the pump with the usual gears and the old magneto chain drove both in the Norton manner.

The engine was fitted into a standard ES2 frame and thus looked like a 1954 model fitted with a 1936 engine, this impression coming from the parallel pushrod tubes and the old style mag chain cover.

On the road 70 mph was reached and vibration no more than usual for a 500 cc single. Economy was naturally to diesel standards so figures of 120 mpg were normal. It was intended that, having proved the idea, the head be remade in aluminium with inclined valves and the barrel replaced by an Alfin one, but no more was heard of this interesting Cornish project.

When the next year's models were announced there was sorrow in the Norton camp, for after nearly half a century the side valve singles were laid to rest and no more were model number 1, the Big 4, or the 16H in the lists. Also dropped was another old friend, the model 18, so one more faithful retainer was gone.

In their place came a pair of machines which revived a prewar model number and kept the final Big 4 engine dimensions alive. These were also machines to appeal to the sidecar man, and for the more traditional of these one model retained the rigid frame. This was the 19R and it used the 82 × 113 mm dimensions and 597 cc capacity of prewar days in an engine also fitted to the 19S which used the swinging fork frame of the ES2.

Both engines were just stretched versions of the 500 and hard to distinguish from it apart from the added height which pushed just about all the rocker box out of sight up into the tank recess. Both had light alloy heads as did the ES2 but the compression ratio of the latter was still no more than 6·8:1. All three machines were fitted with Amal Monobloc carburettors.

On the chassis side the 19R used that of the model 18 but with a 4·00 × 18 in. rear tyre to aid comfort, while the 19S and ES2 shared a common set-up. These two were modified with the fitting of a new dualseat of different shape, boxed in rear number plate with rear reflector, a new shape to the front number plate, altered handlebar with horn button built into the right side, polished fork ends and a smaller petrol tank with round badges and thinner kneegrips.

Motor Cycling road tested a 19S fitted with a Swallow sidecar during 1955 and for this it had the lowered gearing from the 19R plus its fatter rear tyre. It proved to be good for 64 mph and for pulling up tree stumps, so retained the good points of the old side valve plus a little more sparkle. The exhaust note was rather noisy and the spanners provided were unable to shift the factory-tightened nuts of the supposedly qd rear wheel.

Near the end of the year the rigid 600 was dropped, cynics suggested because the stock of old frames had been cleared, and a new smaller one joined the two that were left. This again revived a prewar machine, the model 50, of 71 × 88 mm dimensions and 348 cc capacity. It had an alloy cylinder head, a compression ratio of 7·3:1 and used the same cycle parts as its two bigger brothers in nearly all respects.

There were slight increases in compression

Above **1961 slimline ES2 with narrower tank and quickly detachable seat**

Left **The drive side of the 1961 single with crankshaft mounted alternator, this example fitted with optional rear chaincase** ratio for the ES2 and 19S and both had an improved exhaust cam and a common size of exhaust pipe which thus made that of the 500 larger. Both had the cylinder fin area increased.

All three singles had full width, light alloy hubs fitted to both wheels, the brake size remaining unchanged. Also new was a redesigned oil tank styled to match a combined toolbox and battery carrier in which went the dynamo regulator. A new and larger petrol tank, with kneegrips secured by screws, was fitted along with new taps and fuel lines. The oil lines were also changed to make them more substantial and to give them some flexibility.

The front suspension was made softer and the top fork area altered so that the speedometer went onto the panel on the headlamp shell which was made deeper to accommodate it. The light unit was a new type and had the pilot lamp set in the reflector so the unloved, underslung one went. The dipswitch and horn button were combined in one on the left bar.

The ends of the rear suspension fork were changed to forgings and the speedometer drive altered so the cable came from below the wheel spindle. Larger and deeper mudguards were fitted and a better prop stand, while the rider sat on a new style flat-top dualseat. Finally, a change was made from Girling to Armstrong rear suspension units.

During 1956 the road models were fitted with a new type gearbox known as the AMC box but which retained nearly all the old Norton features. The main change was to the positive stop mechanism which became more compact but the essentials remained as they had been since 1935. The clutch mechanism became a roller and cam lever device that required an adjustor in the clutch pressure plate. This brought two snags in its train, one the need for access to the clutch centre, and the other the loss of the large mushroom on the pushrod which had forced the clutch to lift square and free.

For 1957 there were more detailed changes to all three singles. For each the head was a new alloy casting with more fin area, and the valve wells modified to encourage the airflow across the top of the combustion chamber. The new head had integral pushrod tunnels so the tubes now sealed between it and the crankcase. It also had the inlet flange cast in one with it. Inside were new cams with extended quietening ramps and a little more lift, while the appearance of the mag-dyno was improved.

The frame lost its cradle casting and became all tubular, and the front fork crown was altered to increase the trail. At the rear the fork pivot was narrowed and the torque arm for the brake dispensed with, the backplate locating directly to the fork end. The rear units changed back to Girlings and the chainguard was cleaned up, while the brake pedal return spring was concealed. At the front went a new front hub with the drum cast into the alloy and, while the size was unaltered, a new backplate was used with a forged cam lever in place of the earlier pressing.

The switch and instruments were mounted directly into the headlamp shell so dispensing with the small panel, and plastic grips went onto the handlebars. The silencer was changed to a tubular form with low level inlet and no tailpipe at all. The petrol tank was altered so it only had to be painted, and the side panels complete with badge and kneegrip could be attached by screws and thus plated separately. An optional air filter was offered and, when used, fitted between the oil tank and the toolbox.

The only change for 1958 was to the price, but early in that year Norton lost a major force behind their success when Gilbert Smith resigned his post as managing director after over 40 years' service with the firm. His career with the company had run from office-boy through most departments and in the early days he would on occasion be pulled out of the drawing office to help out with machining or on assembly. In his efforts to sell the make he toured the world and could claim to know every dealer personally at

Not really a Norton at all, the Mk II single introduced for 1965 with AMC engine

home and abroad. He was a keen follower of the TT and from 1920 to his retirement only missed one year when ill. He also attended the Manx and, while actually on holiday, kept a close eye out for any promising talent.

The range for 1959 was down to two ohv models, as the 19S was dropped, its job being taken over by the twins. In fact the twins were by then dominating the marque, although the firm's fame still depended so much on the faithful Manx.

For 1959 the singles followed the twins to the adoption of the alternator and turned to the featherbed frame to make the cycle parts common to the range, while the engines differed only in capacity. The change to the electrics was done by mounting the alternator rotor on the left end of the crankshaft and the stator on a casting spigoted to the crankcase. The chaincase outer was modified to clear the new parts. On the ignition side a coil was adopted and fitted under the rear of the tank and the points went into a unit with auto-advance in the old magneto position and driven by the same chain. The alternator rectifier went under the tooltray under the dualseat.

The frame was the normal featherbed with fittings to suit the single cylinder engine and available in sidecar form with different fork trail and heavy-duty rear suspension units. A full chaincase was another option as was the conventional black and silver finish, for as standard they were supplied in forest green—a major change for the staid singles. Detail alterations were the addition of covers over the engine plates front and rear and, inside, multi-rate valve springs.

The Motor Cycle road tested an ES2 during the year and it was now good for 82 mph and went along the road just like all the singles before it but with even better handling. You could fall off an ES2 but only after chamfering footrest and silencer until you levered the back wheel off the ground. It was not easy to do. The brakes worked very well from all speeds.

Out of the blue Norton received an order for

£20,000 worth of ES2s from Czechoslovakia at the end of 1958, this being their first sale in that quarter for a decade apart from a Manx or two. Several more such orders were to follow in the next three years, all for the same 500 cc single.

1960 brought a change to the internal gearbox ratios of both singles and the option of chrome plated mudguards but there were more extensive changes for 1961. The most useful to the rider was the adoption of the slimline frame used on the twins in the earlier season. This had the upper rear runs of the main loops pulled in to allow the dualseat nose to follow and not splay the rider's legs out so much. The old frame was from then on known as the 'wideline'.

With the new, narrower tank went a quickly detachable dualseat with the tools and electrical equipment beneath it. The tank itself carried a new style badge that encouraged the use of two-tone colours by providing a horizontal dividing line. Options still included the full chaincase, chrome guards and sidecar kit, while flat bars were available in place of the stock items.

Production rolled on at Bracebridge Street despite a small fire at the works early in 1961 but the great days of the company and the singles were drawing to a close. Thus there were no changes to specifications or colour for 1962 and in the following year only to the colour, which became black and off-white for both models. At the end of 1962 Norton production was transferred to Plumstead and the AMC influence grew.

During 1962 it became possible to buy a production racing kit for the two singles, mainly, one suspects, because it was aimed at the twins and fitted all duplex frames which automatically included the singles. The kit comprised rearsets with suitable foot controls and a gear linkage and was fitted by drilling an added hole in each frame gusset plate below and forward of the rear fork pivot.

The faithful singles were finally dropped from the range late in 1963 at a time when the industry was feeling the pinch of a contracting home

market. The accent had turned to big twins for export and customers for the traditional merits of the single were fast vanishing. The last Norton single still retained a line to its engine that could be traced right back to the first Edwardian side valve, and its attributes reflected James Norton's outlook on motorcycling as transport, relaxation, economy and fun.

While this was the end of Norton singles, as known to enthusiasts for the marque the model numbers did reappear late in 1964. By this time

An AJS or Matchless by another name but with Norton forks and wheels so a hybrid alien to the badge on its tank. And it didn't sell

the AMC group was beginning to run into financial problems and for 1965 re-cast its combined ranges of machines into some odd and often unwelcome forms. Thus the Norton twin engine appeared in AMC machines, which upset the Plumstead enthusiasts, while the AMC singles were re-badged as Nortons to the chagrin of those who still mourned for dimensions of 79 × 100 mm.

Two such models were listed as the 50 Mk 2 and ES2 Mk 2 but were Matchless G3 and G80 models with the round plastic Norton badge screwed to the tank. They used the Roadholder forks and Norton wheels and the company also continued to offer the same models with either AMC badge and finish.

For 1966 the tank styling line was altered and Norton transfers added to the side panels, but the group was running on to bigger and more formidable financial rocks, so at the end of the year the singles were discontinued and that brought them to a final finish.

Except for one last prototype, the NVT P92. This was a conglomeration of parts from the factories that had been combined as the remains of the British motorcycle industry in the 1970s and comprised a Triumph Bandit frame modified to use the Isolastic principle from the Norton Commando to hold a BSA 500 cc engine, the B50. Tank and seat were Norton, as were the Commando mute-type silencers fitted to the twin exhaust pipes connected to the single exhaust port. Forks and front hub were both late type BSA-Triumph parts, while the rear hub was lashed up to fit. Some of the detail work, such as the left gear lever and right rear brake, was crude but the machine as a motorcycle ran smoothly and free from vibration at the expense of some low-speed judder. It handled well and would cruise easily at 70 mph, so had many attributes of the Norton single. In 1973 this was not enough as fours were all the rage and by the time singles had become popular once more NVT had gone and so had any chance for the last big single.

5 Wartime

To anyone who kept their eyes open and read the papers the 1939 war did not come as any great surprise; Gilbert Smith did both these things, and had a factory team touring Europe and able to brief him on events and their background.

The money and effort going into the BMW, DKW and NSU racing teams was seen and noted, along with the glider airforce of Germany in that period. Summed up it all indicated a major conflict and that meant the peaceable government and its ministries would overnight want transport of all kinds in vast numbers.

Norton were in the transport business and Gilbert Smith was determined that the company should receive a large share of any orders that appeared. In this he was prompted by the desire to see the Norton company have a lucrative government contract and also, one suspects, because they were good at making motorcycles and he wanted them to go on doing that, not become just another munitions factory turning out anonymous parts. Furthermore, he must have realised that at the end of a war it would be much easier to pick up the manufacture of motorcycles if there was no break in production, as the assembly lines and personnel would all be suited to the task.

To make sure that Norton was well known to the army Gilbert did two things. The first was to call regularly at the appropriate office in Whitehall and ask to sell some motorcycles. This

First of 100,000, the 1937 16H on which the WD model was based minus a few refinements such as the tank top instrument panel

approach through the official front door was met with a steadfast denial that they might need motorcycles, or for that matter anything else, for at that time the government was on a pacifist tack and no public acknowledgement of the impending crisis and its needs was made in speech or by buying equipment.

The second avenue that Gilbert pursued was to cultivate a number of Army officers and arrange for them to attend events such as the TT, the ISDT and the Scottish, where the Norton make could be seen to be performing well and where the officers could themselves be well looked after. This last aspect extended to arranging for reports to be written by a member of the Norton staff to save them that chore.

In the end all this background work paid off and Norton were awarded a contract by the War Office for machines. The one selected was based on the 1937 16H engine with the open valve gear, this being preferred to the then new and untried 1938 version with enclosed valves. The engine went into the trials frame with high ground clearance, the trials crankcase shield was added along with the buffer stops on the forks, and the army had its motorcyle.

The 16H proved to be absolutely right for this

Above **The basic 16H which did such sterling service in so many theatres of war**

Right **Royal Warwickshire reconnaissance unit in training in Ulster with a 16H following a Big 4 sidecar outfit (IWM)**

work, being simple, cheap to build, and easy to keep going. Just what was needed in war where equipment was often abandoned for one reason or another and a throw-away policy forced on the men. During the peaceful years of the middle 1930s the War Office had bought vee twin 500 cc BSAs specially built to a rather elaborate specification and these had been fine when there was leisure and tools to look after them.

After September 1939 there was to be no leisure for a long time and tools were hard to find in the middle of a battle. Then the despatch riders needed to be able to improvise to keep going or to dump the model and take to walking. Most were allergic to the latter course and found the simple side valve with its long history could normally be persuaded home.

Faults there were and ones well known to the motorcyclists in the ranks. The rear hub suffered as always from loose studs, the magneto or its platform could work loose, and the gearchange could wear to the extent that selection became somewhat vague. In the field these points could always be remedied to keep mobile even if not fixed in the normal sense.

So Norton kept on building them right through

B
55
55

C
4399326

Deep in the Dutch mud late in 1944 after days of rain (IWM)

the war until over 100,000 had been delivered to the forces. Nothing of any importance had to be changed and so they could be used, abused, worn out and sent back to depot for rebuild without any spares problems at all. Most went through this procedure at least once and some several times, so nearly all lost their original identity for they were just stripped, the parts repaired as needed and then binned for re-assembly at a later date.

Left **A 16H delivers the** *8th Army News* **to an armoured car in the desert (IWM)**

The changes that did take place were mostly to do with saving cost, materials and production time. The traditional Norton footrests with their little end plates and shaped rubbers were replaced by trials type tubes with a disc on the end to stop the boot sliding off. Handlebar grips

Above **A variation of the Big 4 outfit with a mortar on the sidecar chassis (IWM)**

Right **A Don R with steed marshalling a convoy with pointed suggestions from the right (IWM)**

became WD webbing, and the speedometer moved to one side so that its cable did not get trapped in the forks. The toolbox was fitted with the one winged screw that might slacken but not unscrew and the headlamp size was reduced.

From the one basic machine a number of variants came, but the changes were very minor to suit different theatres of war. Overseas an air cleaner was mounted above the right pillion rest and connected with flexible hose to the carburettor and the tail of the silencer turned up. For the desert the air cleaner could be even larger but all these added items were easily fitted as extras to the basic machine. Colours varied according to the service they went to, with dark and light blue for the navy and RAF, and khaki, camouflage finish or desert sand for the army models, depending on the intended field of operation.

As soon as the 16H contract was in the bag Gilbert Smith started to pester the ministry for another one to develop and build a sidecar outfit for the motorized cavalry. Again he had a headstart over his rivals for Dennis Mansell, son of the Norton managing director at that time, had been using a sidecar wheel drive outfit for some years whenever it was allowed. Thus they had a good idea of the problems likely to occur and their solution.

They got their contract and based their answer on the Big 4 engine which had much in common with the 16H. They used the 1938

version with enclosed valves but the real differences were to the cycle parts. The design used a live rear wheel spindle from which the sidecar drive was taken across to its wheel by shaft with a simple dog clutch operated by a hand lever to connect the drive when needed.

The wheel bearings were thus located in the rear fork ends which were much enlarged to accommodate them. While doing this the frame and forks were modified to take a 4 in. tyre section, including the knobbly trials type, by providing more clearance where needed. A hefty lifting handle and mudguard stay was also added for hauling the outfit out of muddy predicaments.

The sidecar was based on the standard Norton chassis which supported the sidecar wheel on both sides with a long fixed spindle which passed through the outer frame lug and the wheel before screwing into the main chassis. This avoided the normal overhang of the stub axle set-up and gave the wheel a measure of protection plus a firm support for the mudguard.

The sidecar chassis was altered to take the bearings for the live wheel spindle and all three wheels were of the normal Norton three-bolt pattern and interchangeable, with a spare carried on the back of the sparse body. This was simple and stark with the accent on ease of entry and egress plus a mounting point for a machine or Bren gun, rather than creature comforts. The shape was roughly based on the trials chair of the period and the body had grab handles front and rear plus a stowage space behind the passenger.

A second hefty lifting handle and stay went on the front mudguard along with a half enclosed front wheel on some models, done by valancing each side of the mudguard into the wheel spindle. The spare wheel could be mounted on the front face of the chair nose, although this space was normally used for unit marking.

When the outfit went into test it soon became apparent that 634 cc on a modest compression ratio was not really up to the job of hauling the weight or forcing the knobbly tyres through the mud. Traction it had but more urge was needed to take advantage of it.

So the Big 4 was tuned by shifting the combustion chamber volume over the valves

Right **Many WD 16H models were demobbed in this way, bought by the trade and quickly resold**

Below **Workshop rebuild in 1942, a point where engines and frames became separated for always**

BIGGEST BARGAIN OF THE YEAR!
NeR
Norton
REGD TRADE MARK
UNREGISTERED Ex-W.D.
MODEL 16H, 490 S.V.
Actual photo taken in our Showrooms
IMMEDIATE DELIVERY FOR £24·10 DEPOSIT
Once again NAYLOR & ROOT make news with a wonderful bargain offer of unregistered ex-W.D. machines! We now offer 200 of the famous Model 16H NORTON 490 c.c. S.V., with 4-speed footchange, fully equipped and smartly finished, on the following amazingly easy terms — only £24.10.0 deposit and £4.10.0 monthly, or cash £94.10.0. An ideal machine for solo or sidecar use. Here, indeed, is your opportunity!
Note the big features listed below.
★ CHROMED EXHAUST PIPE AND SILENCER
★ CHROME LAMP RIM
★ LUCAS MAGDYNO LIGHTING
★ SPEEDOMETER
★ ELECTRIC HORN
★ NEW BATTERY
★ PILLION SEAT & RESTS
★ PANNIER FRAMES
★ CIVILIAN FINISH
Attractively cellulosed in a pleasing shade of green, with gold name transfers on tank.
Most of these machines have been overhauled in Army Auxiliary Workshops. They are offered by us ready for the road with 3 MONTHS WRITTEN GUARANTEE. To get one of these super bargains, call or write at once.
PER PASSENGER TRAIN TO YOUR NEAREST STN.
Post orders will be dealt with promptly and in strict rotation. Send deposit of £24.10.0 under our guarantee, or, if you prefer, first send the coupon. But, to avoid disappointment, do not delay.
Post Coupon Now!
I am interested in your bargain offer of Ex-W.D. NORTON Model 16H, 490 S.V.
Please send me full details and simple Mail Order Form per return.
NAME
ADDRESS
M.C.137
NAYLOR & ROOT LTD
248-250, LAVENDER HILL, CLAPHAM JUNCTION, S.W.11
Phone BATtersea 5272 (7 lines)
Hours: 9 a.m. to 6 p.m. Wed. 1 p.m.

and introducing minimum head to piston clearance over much of the bore as set out by Ricardo. It really is the only way to get any sort of decent compression ratio with a side valve engine and gives the bonus of good turbulence from the squish action.

One side result of this activity was to alter the head, barrel and piston a little so, although they look much the same, the parts are not as those of the 1938 engine, the barrel and tappets being longer. This change was of no matter to the army as the parts were all different to the 16H anyway and the modification was continued after the war for civilian Big 4s.

Quite a large number of outfits were built and used mainly for reconnaissance work in the early years, many going to North Africa. Despite the special skills needed by driver and crewman, for chair work is very different to solo riding or car driving, they functioned well and proved useful in a number of quarters.

Experience in the field suggested a number of points that could be improved upon so work began on a second version. The exposed drive shaft and dog clutch had suffered undue wear and tear so it was proposed to fit a car-type rear axle carrying both rear and sidecar wheels. Also required was a reverse gear to enable the crew to back out of tricky situations, while the need for more power to haul it through deep sand or mud was answered by a big twin engine from AMC. This was a course taken by the Germans with their sophisticated BMW and Zündapp outfits that went to double gearboxes, reverse gears and lockable differential before the combined problems of weight and manufacture turned them aside.

The Norton outfit lacked some of the special features of the enemy models and the mark II would have been heavy. With no sidecar wheel lead, as a dictate of the axle design, it would also have handled in a strange way even for a chair. And then Lease-Lend came about and with it the Jeep which carried more and could be driven by anyone. In times of stress you could hide under it as well!

In only one respect was the sidecar to be preferred and that was on a skyline where its odd silhouette without a defined profile made it much harder to pick out than the normal foursquare box of the four-wheeled vehicle. Of course a really skilled chair driver and passenger could get through situations the Jeep might cavil at, but they were rare birds and in average hands the outfit could become a mobile disaster.

The Army thanked its stars for the Jeep, forgot sidecar outfits, and kept its solos in the style of the 16H with lightweights for training and other jobs.

Early in the war the War Office asked makers for prototypes of machines more suited to cross country work than the modified civilian machines they were then using. All of them, the 16H, BSA M20, Enfield, AMC, Triumph and Velocette were far too heavy for the tasks they were having to perform so the aim was to reduce weight, have enough power to cope, keep it simple and leave plenty of clearance under the crankcase.

The Norton answer was a 350 cc side valve engine in a welded frame with light wheels. To get the weight down much use was made of light alloy with the head in this material and also the barrel, which had cast-in valve seats and liner. The timing cover was electron and even the handlebar levers were aluminium. The timing chest differed from the usual Norton single with a single camwheel and bellcrank rockers to work the valves.

Norton were given a contract to build 50 of these machines for trials in service use and nearly all went to France and were abandoned in the retreat to Dunkirk. After that, matters were rather grim so all such experiments were forgotten and the factory just had to get on with building the 16H as quickly as possible.

This they did until the war ended and the Ministry found itself with vast numbers of unwanted machines on its hands. At first these

Most finished their days like this, certainly mine did. But this is actually an early postwar model fitted with an ex-WD engine

were stored in open air dumps, but in time and with some prodding from a transport hungry public they were sold off. Not as single units but auctioned in big lots to those dealers with the enterprise and money to buy.

Often these machines were fully overhauled by the army workshops before release, as such repair work kept otherwise idle hands busy until they were demobbed. They might be repainted but only in khaki or an army green so usually the dealer would give the machine a quick coat of paint in a colour less associated with the services, often maroon but fortunately more often black for the Norton.

A number of the sidecar outfits also reached the market but all had the sidecar drive cut off by flame to ensure no comeback with accident claims. The army knew, even if possible civilian owners did not, that if you forgot and kept the sidecar drive engaged when you left the rough and went back to tarmac the outfit would refuse to turn corners. With heavy tyres and no differential it just went straight on.

Most of the military machines finished their days in mundane fashion, often towing a chair and finally being broken for spares or discarded when their owners could afford something better. For a brief time they filled a very real gap and helped many onto wheels when otherwise they would have walked. Now they are collector's pieces.

6 500T and off road

The Norton range of machines had offered something for the trials rider for many years before the war by dint of a Colonial specification. This practice went back to the 1920s and was then done as a means of making any machine more suitable for overseas and developing countries. The only real changes were to the frame, which had more ground clearance, and the exhaust system which ran at a higher level.

The alternative specification was available for all models so Norton were able to offer the complete range from mundane sidevalve to exotic ohc to all whose roads were often unsurfaced paths. This practice was continued throughout the 1930s so, in theory at least, it was possible to order an International with optional alloy engine and to trials specification and with full lighting equipment.

Norton themselves did use just this form of build for their own riders in trials and ISDT events and many of these did fit the camshaft engine in 500 and 596 forms, the latter in the sidecar class.

After the war Norton tried the same arrangement to produce 350 and 500 Trials models for 1947. At that time they had no 350 ohv engine at hand but for the larger model took the iron model 18 motor and installed this in the WD 16H trials type frame fitted with the new Roadholder forks. A high level exhaust was used but that aside most of the cycle parts were from the road model and were fitted. Thus it emerged with dynamo, battery, lights, electric horn, unsprung

Below **The ohv single to trials specification in 1938 with competition tyres and mudguards**

Bottom **1939 model 18 in trials form. More acceptable silencer than earlier machine**

and close fitting front mudguard and with the standard tank finish. It was given narrow section mudguards and the tail of the rear one was abbreviated a little. The gearing and tyres were made more suitable but even for the trials of 1947 the machine was gross, being too heavy, wide, low and long. Private owners changed to other tanks and girder forks, there being plenty of those around, and this reduced the wheelbase to a more acceptable figure for trials work.

The Trials model was not listed for 1948 but in the preceding winter and through the early part of the year works prototypes were seen. These were the fore-runners of the 500T that was to be added to the range late in the year.

The wheelbase of the new model was reduced by shortening the rear stays and fitting new yokes that brought the fork tubes back in line with the steering column. The frame remained the faithful diamond type 16H to which was bolted a crankcase shield under the engine, and this became a very tight fit indeed. So much so that the rocker box could not be removed without first taking the cylinder head off. The gearbox was the old prewar style tucked into the small space left for it between the back of the crankcase and the rear wheel. It was fitted with wide ratios and a folding

Right **Early 500T from 1949 with special fork yokes and scolloped front hub flange**

Below **A prewar ISDT machine with racing magneto, air bottle, spare fork spring and Union Jack on tank**

Norton
JOK 130

kickstart lever but had the boss for this item reduced in size and more tucked in.

The engine was basically the normal 500 but fitted with a bi-metal alloy head and linered alloy barrel. Internally cams and flywheels were standard ES2 while the compression ratio was 6:1. The exhaust swept down to be clipped to the front engine cradle and then over the right footrest to an uptilted silencer. The silencer was of special light construction and a good few other parts were aimed at keeping the weight low. Among these were alloy rear engine plates, number plates and mudguards, but more fundamental was the fitting of a 7 in. front brake with a water excluding ring in a scalloped hub and a special alloy brake plate with ribbed support for the cam spindle boss. To lighten the cast iron brake drum derived from the roadsters, large holes were drilled and then blanked off with core plugs.

The petrol tank was welded up in a light steel and held 2·5 gallons. It was mounted on to a pair of horizontal, rubber insulated prongs at the front and held down on to rubber by a single fixing at the rear so was quick and easy to remove. The oil tank was also lighter than usual and had its filler cap set in the top to keep it well tucked in and out of the rider's way.

Wheel rims were in high tensile steel so lighter than usual and carried trials tyres. The saddle was also lightweight, being a rubber Dunlop item, and had alternative mountings to enable the seat height to be varied. The handlebars were adjustable and the control levers simple pressings. A toolbox was fitted.

Lighting was quoted as optional with a quickly detachable headlamp with plug and socket wiring to it. Prop and rear stands were supplied and the finish was black with the tank dull chromed. Ignition was by a BTH magneto and among the good detail points was the way the standard rear brake pedal was tucked well in.

It was a successful trials machine and popular with private owners. The works models were little different, although as time went on they did gain

Right **Trials Norton built by Francis Beart around 1948 using mid-1930s engine**

Below **1949 500T showing the special fork position and the tight fit of the engine**

The 500T in 1951, no scollops but otherwise little altered

more alloy items such as chaincases and prop stand legs among others plus the lighter Manx clutch, but nothing an enterprising owner could not manage for himself.

The ISDT machines were rather more special and for that event had plunger rear suspension, a Manx front brake and a TT carburettor plus all the usual fittings to speed up repair work and enable any wiped off controls to be rapidly replaced.

A further variant used by Geoff Duke had a pair of model 19 flywheels fitted to raise the capacity and, to allow the normal alloy barrel to be used, a large spacer under the barrel. To get it into the frame it had to be set lower and the ground clearance reduced. The massive flywheels allowed the bike to plonk even better than the 500 which suited the sections of those days.

The only changes made for 1950 were the adoption of the one-piece oil junction block common to the road models and the raising of bottom gear. By 1951 the scollops had gone from the front hub rim but the weight was still only 300 lb, thanks to the alloy engine with Alfin barrel and head with valve inserts in the aluminium. A prop stand was still provided attached to the left lower chainstay but, unlike the road models, the 500T had one with a large toe. The standard, vintage style rear stand was usually discarded by owners.

There were rumours of a 350 cc version being added to the lists but officially this never came about and it does not appear in contemporary literature. While the works might have built one or two specially to capture the capacity cup in the trials they entered, there seems to be no record of them doing this and riders from those days don't recall seeing the machines at all. While it would have been easy for the competition shop to have prepared such an engine from spares, the likelihood of a production model is remote as at that time the firm had no 350 ohv machine in its lists, the revived model 50 not reappearing until late 1955.

There were no changes for 1952 but the following year the power quoted went up a little to 24 bhp at 5300 rpm, possibly due to the reported use of a die-cast cylinder head. For 1954 it was suggested that a Lucas magneto had taken over from the BTH but otherwise the 500T continued in its original form. In September it went from the range and was not replaced as Norton never listed a trials model again.

While the factory did have some involvement with the ISDT in the early postwar years, the singles used were modified standard ones using the 500T as a basis in most cases but with parts as thought best from the road and racing models. Their main interest, outside road racing, was trials and it was for such events that they recruited Geoff Duke at the start of 1948 from BSA after he had received an offer from AMC. He won the Northern Experts that year and continued to ride in trials up to 1951 when he was seen in the Colmore on a featherbed trials model with the engine oil carried in the frame tubes. This was a typical works lash up and the wide set of the frame tubes below the engine and at the seat cannot have been much help. The tank was from the 88 twin and a dualseat was used with an air filter under it. The rear suspension units were set nearly vertically but did not have remote reservoirs as used on the racers.

Norton also did well in sidecar trials in that era with Arthur Humphries being very successful and winning the Experts using an old type longer wheelbase frame driven by a 500T alloy engine. He kept to the solo gearing and had 3·00 × 21 in. tyres on front and sidecar wheels, the latter fitted with a brake operated by a hand lever on the grabrail. The front wheel had a conical hub and Manx backplate.

In 1952 Norton prepared four scramblers using a combination of racing and trials parts. The frame type used was basic 500T but modified to swinging fork rear suspension with rear units longer than the racing ones. Into these frames

The featherbed scrambler of 1952 based on the Manx crankcase and Inter alloy head in a modified 500T frame

Norton

went either 350 or 500 cc camshaft engines based on the Manx with the racing crankcase with four studs along its base. A Manx barrel was used but the head and cambox both came from the International, the head being in light alloy.

Petrol and oil tanks, seat pan and mudguards were in the same material and fabric-lined guides kept the rear chain on the sprockets. The gearbox used was the laid down type and engine plates were in alloy. The front forks looked standard but had some Manx parts in them, and a racing brake was used in the 21 in. front wheel.

After this exercise Nortons left the scrambles field alone but Les Archer kept their name very much to the fore with a series of camshaft engined machines. These used frames with duplex downtubes but single top so, although they seemed much like the featherbed, they were all special. Norton forks were used but the brakes were Manx and in conical hubs. There were a good number of special or modified parts in the end as the machine was altered in the light of experience. Thus the chain line was set out to clear the knobbly rear tyre, while the primary chain had a standard Norton chaincase adapted to fit to the Manx crankcase. The tank came from a 500T and was reduced in volume, while the oil tank was very special. It was made like a house with two stories with small bore tube connections between the two to prevent aeration and to aid cooling.

Les won the European title in 1956 in a decisive manner with four wins, a 2nd and a 4th from his rides. He was still riding in Europe in 1965 and by then the engine, still a Norton and prepared by Ray Petty, had a twin cam head, short stroke needle race vertical shaft and all the other useful parts. The barrel was a Mahle with hard chrome direct on the bore and fairly sparse fins that needed little cutaway for the cam drive tube. The head gave an 11:1 ratio and carried twin plugs

Left **Special built by Jack Wright for SSDT using part featherbed, part 500T frame, 500T forks and engine, twin gearbox and ingenuity**

but was well pruned of fins as these were not needed and tended to stick into the rider's legs if not modified. The valve springs were changed to the coil type and enclosed, cams were selected to suit scrambles use, and the carburettor was a GP2. Most of the crankcase finning was machined off.

The frame was still mainly duplex but now carried the engine oil, while the front forks were Ceriani. The conical Manx hubs remained along with their brakes. All told it was a very suitable machine for the moto-cross circuit in the mid-1960s and Les normally finished well whenever he rode.

The Norton single, in camshaft form, was also used offroad with three wheels as well as two. For scrambles Bill Howard fitted a double knocker long-stroke 500 engine into an outfit first used by Arthur Humphries in the ISDT with a 350 Inter engine installed. It was basically a featherbed frame in 531 tubing but altered to suit the rigours of the twin trials of a chair and a six days event. The engine also came from a chair but a road racing one and was standard apart from a change of valve timing to suit the straight-through pipe fitted and a reduction in compression ratio to spare the valves if a gearchange was missed.

On grass Bill Evans used a camshaft engine to take him to the British sidecar grasstrack championship six times. His engine finally reached a capacity of 636 cc by using a 90 mm bore top half with the older long-stroke crankshaft of 100 mm. There were even thoughts of finding a 113 mm crank from a 596 cammy engine which would have pushed the size to over 700 cc. Twin plugs were fitted fired by a Lucas twin spark magneto and a big GP carburettor used. On the exit side went a reverse cone megaphone and the engine was good for 7200 rpm, while remaining very flexible.

Thus private owners showed just how versatile the Norton single could be in all its forms and for any competition off road.

7 Racing and records

The Carroll engine and featherbed frame were the final outcome of the racing Norton single which stretched back through girder forks, the Moore engine and the ohv, to the early belt-driven side valve BS.

Their early racing successes came at Brooklands in the days just before the first world war and after it they continued to have some success with the side valve engine. It was, however, the ohv that brought them their second TT success in 1924 after two near misses the year before. That engine was used for two further seasons and in 1926 Stanley Woods won the Senior on one to start his long and very successful association with the firm.

In 1927 came the Moore camshaft engine and Alec Bennett took it to victory in the Senior, while Woods won three major events in Europe. This, like 1924, was an early indication of what the Bracebridge Street concern could do on its day, but they were to struggle through three more years before they reached their golden age.

1930 brought the Carroll engine and some indications of the success to come in that decade, but that year the TT belonged to Rudge. As the season progressed the engine and the team managed by Joe Craig began to settle down, and Woods won in Ulster and France.

From then on for seven glorious years the black and silver team from Birmingham dominated road racing. It was a period during which the machines were developed and refined year

Bill Lacey astride his Norton with Tommy Atkins. A Brooklands picture but probably the 110.8 mph machine plus silencer

by year, always a little better and more reliable than the opposition. To do this for a season is an achievement but to hold off all comers for such a period was an extraordinary feat and as much credit for this went to Joe Craig as to his team of superb riders.

Joe was a development engineer par excellence who honed and refined his machines to get that little more from them. This was done by countless experiments with cams, ports and valves to improve the breathing and combustion. It is said that when Joe was watching an engine running on the test bed it was just an outer shell standing there, the real Joe was inside the combustion chamber watching the mixture flow, the flame front and the exhaust gases.

The Norton race team of the 1930s, its equipment and operation would have been laughable if it had not been for its success. Most of the time it comprised Joe, a pair of each size of machine for the two riders, a mechanic, and a bag of tools! Often they would travel about Europe by train, for this was a fast and easy way to cover the large distances in the days before high speed roads and high speed vans. It also allowed them to rest between meetings and, if some work had to be done, the mechanic could always go in the guardsvan with the bikes and get on with it.

The problem of transport at the meeting, getting from the station to the circuit and to the hotel or digs was dealt with by simply sending a wire to the local agent telling him to get it all organised. Race spares could also be sent out to the same agent, so if any became needed due to a fall or a mechanical fault at one meeting they would be waiting the team at the next. The use of the agent's workshop would be automatic but failing that, the kitchen table at the digs might be pressed into use!

This simple system worked wonderfully well for the whole of the 1930s, and up to 1937 their domination of the racing scene was nearly total. It was when a Norton failed to win that it became news.

In the Isle of Man they began their years of

Left **The Maestro, Duke in the 1950 Junior TT, the day he had to give best to Artie Bell**

Bottom **Eric Oliver with the record breaking outfit at Montlhèry**

Below **P H Carter, winner of the Senior Clubman's TT in 1950 with his Inter. No silencer or dynamo was normal practice**

success with doubles by Tim Hunt in 1931, Stanley Woods in 1932 and 1933, and Jimmy Guthrie in 1934. Guthrie came within four seconds of repeating the double-double in 1935 for he won the Junior but lost the Senior to Woods on a Guzzi in the famous event, deferred one day due to the weather. Guthrie started as number 1, the privilege of the previous year's winner and Woods drew 30 so was riding about half a lap in arrears. Thus Guthrie, leading comfortably on his last lap, was signalled 'ease' at Ramsey just before Woods was due at the end of his sixth lap when he was expected to refuel, and for which his pit was preparing. History recalls that Jimmy eased off, Stanley went straight through and pulled back the 26 second deficit plus four more to win. For once the normally impeccable team management of Joe Craig had been out-manoeuvred.

Guthrie made no mistakes the next year and kept his chin on the tank to the end to win the Senior, while Frith took the Junior, and in 1937 they reversed the arrangement. Through that summer the two Norton men went round Europe winning events and Guthrie became Junior and Senior Champion. The opposition was, however, at last making their lives more difficult and they had to contend with the blown BMW in the 500 cc class and the similarly charged NSU twin in the 350. Good handling, reliability and superb riders saved the day that year, but for Norton the season ended in tragedy when Guthrie died following a crash a mile from the finish of the German.

In 1937 the BMW and NSU teams had been well prepared and the next year they were bigger with teams of mechanics, workshop lorries, spare machines and back-up personnel. BMW had their own physical trainer to monitor food, drink

Right **Cyril Smith at Cadwell Park in 1954**

Below **Bill Boddice at Crystal Palace in 1955**

30

Steve Lancefield and Jim Russell inspect a Manx in a Formula 3 single seater

and exercises, all of which contrasted sharply with the small Norton team, itself with far more resources available to it than any other from England.

1938 saw the introduction of a European Championship based on eight meetings and that year the Norton star began to wane before the power of the blown BMW in the 500 cc class. In the 350 it was the Velocette that gave the problems, not the NSU, and the Hall Green models proved too quick for Bracebridge Street that year and the next. They also had a 500 and in the Senior TT ran the Norton very close but Harold Daniell pulled out the stops on the last lap to win by a small margin from Woods and raise the lap record to a long-standing 91 mph.

Elsewhere the Norton was less successful and Daniell finished 2nd to Meier in the 500 cc championship. 1939 saw Woods hold Daniell down to 2nd in the Junior TT, while the BMWs kept Frith in 3rd in the Senior. They also had to face the blown Gilera which was even quicker.

After the war Joe Craig brushed the dust from the works machines, dropped the compression ratios and the 500 continued to win the TT, but in 1949 only because the AJS stopped three miles from the finish. Elsewhere the old design was showing its age and in the 350 cc class the Velocette was faster. The Junior class always was more of a struggle for the Norton and their machines of that size less competitive than the 500. Weight must have played a part in this and always the 350 was the 500 fitted with a smaller engine and suffering because of it throughout the life of the Carroll motor.

During the postwar period the Clubmans races began in the Isle of Man and the International Norton proved very suitable for this, winning the two major classes in the first event held in 1947. They were out of luck the next year, but in 1949 Geoff Duke began his climb to fame by winning the Senior. Geoff returned to the Island later that year to win the Senior Manx and so was virtually an automatic selection for the works team in 1950, when his talent and the arrival of the featherbed frame gave the Norton a new lease of life. He also brought in one-piece leathers to reduce wind resistance, and the combination was more than a match for the competition.

He should have won a world title in 1950 when he took the Senior TT and broke Daniell's prewar lap record, but tyre failures in two races cost him dearly. He won in Ulster and again at Monza against the full Gilera team but lost the title by a point. In 1951 he took both 350 and 500 cc titles, retaining the smaller class in 1952 but losing the 500 one following a crash in a minor meeting. It was also obvious that he was only getting the results by riding over the limit so he went to Gilera who had been chasing him for over a year.

The Norton should have been finished then but Ray Amm did a TT double on them in 1953 and won a controversial Senior in 1954. He did the double in Ulster but from then on classic wins became rare for the Norton marque. One came in 1958 in Sweden when Duke, near the end of his career, had a winning day, but victories such as Hartle's win in the 1960 Ulster came because the MV had troubles, not from superiority.

There was to be one more glorious double and it is fitting that it came at the TT for the name

Left **A 350 Manx with homemade full dustbin and engine undershield in 1955**

Below **John Surtees on his 350 at the Hutchinson 100 in 1957 at Silverstone. He won this and two others on MVs**

Above **Bob Anderson ahead of Derek Minter at Brands Hatch in 1958**

Norton was synonymous with that event where they had so much success and had won so many races. It came in 1961 when the great Mike Hailwood won three of the four solo events and was leading the 350 when his engine failed on the last lap, which let Phil Read in to win on his Norton.

In the Senior Mike had a very good engine built up for him by Bill Lacey who had worked on his engines before, although that arrangement had come to an abrupt halt. For 1961 Stan Hailwood persuaded Bill to put together one unit to give Mike the best possible chance against the MV of Hocking. It was an engine built from parts, mainly from 1958 and 1961 motors and on the bench it gave 53 bhp. Mike tried it in a race at Castle Combe and Bill let him do a lap or two in practice in the Island. In the race the Lacey engine allowed Mike to pressure Hocking to overstrain the four and bring the Norton home to a famous victory at an average of over 100 mph.

From then on there may have been little for

the Norton at grand prix level, but elsewhere the picture was different. Entry lists bulged with Nortons just as they had since the war and before it. The Manx was the mainstay of road racing all over the world and remained thus until the advent of the 351 cc Yamaha. There were other machines, mainly the 7R, the G50 and, in earlier days, the Mark VIII. None was the dominant force that the Manx was, and at all levels of racing the Norton was the machine to own and the one to beat.

On three wheels Norton had one ineffable weapon in the person of Eric Oliver. In theory the Italian fours should have had no trouble keeping ahead of the single cylinder Norton. In practice it was the reverse or, if they did get ahead, it was only to the last corner when the big banger would sweep past at an impossible speed to take the flag.

Oliver won innumerable events at home and

Below **Bob McIntyre cornering his 350 Norton in his own magnificent style**

Norton

Above **Mike Hailwood on the Ecurie Sportive 350 at Aberdare in 1958**

Left **A very typical racing Manx picture with Bob Anderson on a 350 in 1958**

abroad and in six years of chasing the world title won it four times. In that series he won 17 of the 24 races he started in, and in the first three years was only held back to 2nd once. When he finished lower or retired it was due to machine problems and it was only two accidents in minor events that prevented him making a clean sweep of the titles.

For Norton, Cyril Smith stepped into the breach in 1952 to take the title but in 1954 it went to Noll and BMW, to begin that make's long domination of the class. That year Oliver did win his TT and later generations saw him again on the big Norton at vintage meetings when, a retired pensioner, he lacked none of the style and verve of 30 years before.

Across the Atlantic the Norton single had a run of success on the Daytona combined road and beach circuit comprising two miles of each joined by banked turns at each end. A Norton won the 1941 Experts event and in 1948 Gilbert Smith arranged for Steve Lancefield to go out and look after the two Manx models which had been prepared by the works to suit the American rules. These called for kickstarters and low compression ratios more suited to their domestic side valve vee-twins.

After a confused race one of the Nortons was second but Steve was unable to go out again the following year so passed his experiences on to Francis Beart. Beart was retained by the works for the event and prepared for it in his usual meticulous fashion. His work paid off and Nortons won both the Experts and Amateur

Lovely Isle of Man picture showing a Norton approaching Creg-ny-Baa and the drop down from Kate's Cottage

events. He went out twice more with specially prepared Manxes, winning the Experts each time and the Amateur in 1951. For 1952 the home men were left without works assistance and not allowed to use the featherbed frame model, but still finished 1st and 2nd to bring the marque's score to four wins in a row.

Even in the field of endurance racing the Manx Norton left its mark for the Bol d'Or was won by Lefevre seven times in the years after the war and any machine that completes a 24-hour race has to be both fast and reliable.

It was such traits that led to the adoption of the Manx engine in 500 cc Formula 3 cars. This

class had begun in 1946 to provide low cost racing and in its early days the JAP speedway engine was popular. As chassis improved more power and speed were wanted and the Norton engine was a natural for the job. The days of low cost soon went and in the 1950s drivers would buy a complete Manx just to get the engine and gearbox from it. In the end spectator interest waned and the class fell into disuse but not before some good men had made their names.

The characteristics of a good racing engine can be easily turned to record-breaking, especially over the longer distances and times where the stamina of both rider and machine is all important. Norton were involved in such activities from their very early days and nearly always at Brooklands where O'Donovan, Horsman, Denly and Lacey were among the many who set records on Norton machines.

Of all the records over any distance one stood out as presenting the greatest combined challenge to man and machine—the one hour. It perhaps lacked the need for long suffering physical endurance that made all the really long distance figures such a grind, but 60 minutes was long enough to bring cramp and boredom, while the speed was high enough to make the venue a problem.

Horsman was the first to take the Hour on a Norton covering 71·68 miles in 1920 using a chain drive side valve. Bert Denly followed to push the figure on several times reaching 87·07 in 1924. His final run was at Montlhèry in 1927 when he raised the 500 cc record to 100·57 miles, the first time three figures had been reached by a half-litre machine. Back in England, Bill Lacey on a Grindlay-Peerless JAP became the only rider to break the hour at over the 100 mark at Brooklands with its bumps and call for a silencer. He did this in 1928 with 103·3 miles.

For his next attempt he used the Carroll engine and Montlhèry track. The year was 1931 and, after a mix-up with signals, he circulated with Bert Denly, who held the record, giving him the necessary information. He covered 110·8 miles and Norton once more held the record, often called the Blue Riband of speed.

They raised it some more in 1935 when Jimmy Guthrie used an alcohol burning Joe Craig motor to cover 114°09 miles at the same circuit. It took two visits to gain that record for on the first try in 1934 they ran into all manner of problems culminating with first one and then the spare engine blowing up in the fullest sense of the word when the gudgeon pin broke. On that occasion they did bag a couple of short distance records and found that lap speeds of 116 were possible.

In 1937 the record went to Gilera and in 1939 this machine raised it again to 127·5 miles, where it stayed until 1953. Late that year Norton took a team of riders to Montlhèry for record-breaking. This was not their first such visit for there had been one late in 1949 when 21 records in 350, 500 and over solo, and 600 and over sidecar, had been taken. Also, Geoff Duke found that Artie Bell was faster because his two-piece leathers were a better fit, which led to the one-piece racing suit. Eric Oliver tried hard to raise the sidecar Hour record to over the 100 mark but two attempts were brought to an abrupt halt by flat and worn out rear tyres. In the end he had to curtail his speed to 95·89 miles. A year later he had it up to 97·04 miles.

For the 1953 visit Norton took the silver fish model used that year in the North West 200 and TT practice with Ray Amm and Eric Oliver to do the riding. They started with a 350 cc engine fitted and took 43 records then changed to the 500 and gathered a further 18, including the very important Hour at 133·71 miles with Amm aboard. During the attempt the machine ran minus its tail except for the 10 miles distance on the 500 and during this run Amm covered one lap at 145·41 mph, a record for the track.

It was the last time a single was to hold the record and fitting that it should go to the marque associated more with the racing and record-breaking single than any other—the Norton.

START
STOP
ENGINES
21
21
COUGH
F. WOOD

Appendix

Specifications

Model	**30M**	**40M**	**30M**	**40M**
Year from	**1946**	**1946**	**1951**	**1951**
Year to	**1950**	**1950**	**1953**	**1953**
Bore (mm)	79·62	71	79·62	71
Stroke (mm)	100	88	100	88
Capacity (cc)	498	348	498	348
Compression ratio (to 1)	7·23	7·33	8·67 **1**	8·96 **1**
Valve position	ohc **2**	ohc **2**	dohc	dohc
inlet opens BTDC	60 **3**	60 **3**	60	60
inlet closes ABDC	67·5	67·5	67·5	67·5
exhaust opens BBDC	85	85	85	85
exhaust closes ATDC	45	45	45	45
Valve clear. (cold) in. (in.)	0·012 **3**	0·012 **3**	0·012	0·012
Valve clear. (cold) ex. (in.)	0·024 **3**	0·024 **3**	0·024	0·024
Ignition timing degree			36	36
Points gap (in.)			0·012	0·012
Box ratio: top	1·0	1·0	1·0	1·0
Box ratio: 3rd	1·1	1·1	1·1	1·1
Box ratio: 2nd	1·33	1·33	1·33	1·33
Box ratio: 1st	1·77	1·77	1·77	1·77
O/A ratio: top	4·42	5·16	4·64	5·16
O/A ratio: 3rd	4·86	5·67	5·1	5·67
O/A ratio: 2nd	5·86	6·85	6·18	6·85
O/A ratio: 1st	7·82	9·14	8·22	9·14
Front tyre (in.)	3·00 × 21	3·00 × 21	3·00 × 19	3·00 × 19
Rear tyre (in.)	3·50 × 20	3·25 × 20	3·25 × 19	3·25 × 19
Rim front			LA	LA
Rim rear			LA	LA
Brake front dia. (in.)	8	8	8	8
Brake front width (in.)	1·5	1·5		
Brake rear dia. (in.			7	7
Front suspension	teles	teles	teles	teles
Front movement (in.)	5·7	5·7		
Rear type	plunger	plunger	s/a	s/a

Model	30M	40M	30M	40M
Year from	**1946**	**1946**	**1951**	**1951**
Year to	**1950**	**1950**	**1953**	**1953**
Petrol tank (Imp. gal)	4·5	4·5	5·25	5·25
Oil tank (Imp. pint)	8	8	8	8
Ignition system	magneto	magneto	magneto	magneto
Wheelbase (in.)	54·5	54·5	56	56
Ground clear. (in.)	5·5	5·5		
Seat height (in.)	30	30	30	30
Width (in.)			28	28
Length (in.)			82	82
Dry weight (lb)			298	298
Power: bhp	34·5 **4**	27 **4**	37·5	29·5
@ rpm			6200	6800

1 1951—13:1 option. **2** 1949—dohc. **3** 1949/50 engine. **4** in 1950.

Model	30M	40M
Year from	**1954**	**1954**
Year to	**1963**	**1963**
Bore (mm)	86	76
Stroke (mm)	85·8 **1**	76·85 **2**
Capacity (cc)	498 **1**	349 **2**
Compression ratio (to 1)	9·53 **3**	9·72 **4**
Valve position	dohc	dohc
inlet opens BTDC	74 **5**	82 **6**
inlet closes ABDC	94	95
exhaust opens BBDC	72	94
exhaust closes ATDC	64	70
Valve clear. (cold) in. (in.)	0·014	0·014
Valve clear. (cold) ex. (in.)	0·028	0·028
Ignition timing degree	35	40
Points gap (in.)	0·012	0·012
Box ratio: top	1·0 **7**	1·0 **7**
Box ratio: 3rd	1·1	1·1
Box ratio: 2nd	1·33	1·33
Box ratio: 1st	1·77	1·77
O/A ratio: top	4·4 **8**	5·4 **9**
O/A ratio: 3rd	4·86	5·94
O/A ratio: 2nd	5·9	7·19
O/A ratio: 1st	7·8	9·56
Front tyre (in.)	3·00 × 19	3·00 × 19
Rear tyre (in.)	3·50 × 19	3·50 × 19
Rim front	LA	LA
Rim rear	LA	LA
Brake front dia. (in.)	8 tls **10**	8 tls **10**
Brake front width (in.)	1·5 **11**	1·5 **11**
Brake rear dia. (in.)	7	7

Model	30M	40M
Year from	**1954**	**1954**
Year to	**1963**	**1963**
Brake rear width (in.)	1·75	1·75
Front suspension	teles	teles
Rear type	s/a	s/a
Petrol tank (Imp. gal)	5·25	5·25
Oil tank (Imp. pint)	8	8
Ignition system	magneto	magneto
Dry weight (lb)	298 **12**	298 **13**
Power: bhp	47 **14**	35 **14**
@ rpm	6500	7200

1 1958—85·62 mm/497 cc. **2** 1958—76·7 mm/348 cc. **3** 1958—9·75. **4** 1958—100. **5** 1956—70/100/82/64. **6** 1956—74/85/89/70. **7** 1958—AMC box—1, 1·1, 1·332, 1·78. **8** 1962—4·23, 4·65, 5·63, 7·52. **9** 1962—5·12, 5·64, 6·81, 9·11. **10** 1962—dual 7 tls. **11** 1962—1·625. **12** 1962—313. **13** 1962—307. **14** in 1956.

Model	CS1	CJ	30	40
Year from	**1932**	**1932**	**1932**	**1932**
Year to	**1939**	**1939**	**1939**	**1939**
Bore (mm)	79	71	79	71
Stroke (mm)	100	88	100	88
Capacity (cc)	490	348	490	348
Compression ratio (to 1)	7·0	7·5	7·5	7·75
Valve position	ohc	ohc	ohc	ohc
inlet opens BTDC	35	35	47	47
exhaust closes ATDC	40	40	43	43
Valve clear. (cold) in. (in.)	0·006	0·006	0·010	0·010
Valve clear. (cold) ex. (in.)	0·010	0·010	0·020	0·020
Ignition timing	0·625 in.	0·687 in.	52°	52°
Points gap (in.)	0·012	0·012	0·012	0·012
Front tyre (in.)	3·25 × 26	3·25 × 26	3·00 × 27	3·00 × 27
Rear tyre (in.)	3·25 × 26	3·25 × 26	3·25 × 27	3·25 × 27
Brake front dia. (in.)	7	7	7	7
Brake front width (in.)	1·25	1·25	1·25	1·25
Brake rear dia. (in.)	7	7	7	7
Brake rear width (in.)	1·25	1·25	1·25	1·25
Front suspension	girder	girder	girder	girder
Rear type	rigid	rigid	rigid **1**	rigid **1**
Petrol tank (Imp. gal)	2·75 **2**	2·75 **2**	3·75	3·5
Oil tank (Imp. pint)	3·5	3·5	6	6
Ignition system	magneto	magneto	magneto	magneto
Generator type	dynamo	dynamo	dynamo	dynamo
Battery (Volt)	6	6	6	6
Wheelbase (in.)	54·75	54	54·75	54
Ground clear. (in.)	4·25	4·25	3·75	3·75
Seat height (in.)	27·5	27·5	28·5	28·5

Model	**CS1**	**CJ**	**30**	**40**
Year from	**1932**	**1932**	**1932**	**1932**
Year to	**1939**	**1939**	**1939**	**1939**
Width (in.)	30	30	29	29
Length (in.)	84·75	84	84·75	84

1 1938—plunger option. **2** 1935—3·25.

Model	**30**	**40**	**30**	**40**
Year from	**1946**	**1946**	**1953**	**1953**
Year to	**1952**	**1952**	**1958**	**1958**
Bore (mm)	79	71	79	71
Stroke (mm)	100	88	100	88
Capacity (cc)	490	348	490	348
Compression ratio (to 1)	7·23 **1**	7·33	7·2	7·3
Valve position	ohc	ohc	ohc	ohc
inlet opens BTDC	47·5	47·5	47·5	47·5
inlet closes ABDC	70	70	70	70
exhaust opens BBDC	85	85	85	85
exhaust closes ATDC	42·5	42·5	42·5	42·5
Valve clear. (cold) in. (in.)	0·010	0·010	0·010	0·010
Valve clear. (cold) ex. (in.)	0·020	0·020	0·020	0·020
Ignition timing degree	42·5	42·5	42·5	42·5
Points gap (in.)	0·012	0·012	0·012	0·012
Front tyre (in.)	3·00 × 21	3·00 × 21	3·00 × 19	3·00 × 19
Rear tyre (in.)	3·25 × 20 **2**	3·25 × 20	3·25 × 19	3·25 × 19
Rim front	WM1	WM1	WM2	WM2
Rim rear	WM2	WM2	WM2	WM2
Brake front dia. (in.)	7	7	8	8
Brake front width (in.)	1·25	1·25	1·25	1·25
Brake rear dia. (in.)	7	7	7	7
Brake rear width (in.)	1·25	1·25		
Front suspension	teles	teles	teles	teles
Front movement (in.)	5·7	5·7		
Rear type	plunger	plunger	s/a	s/a
Petrol tank (Imp. gal)	3·75	3·75	3·75	3·75
Oil tank (Imp. pint)	6	6	7	7
Ignition system	magneto	magneto	magneto	magneto
Generator type	dynamo	dynamo	dynamo	dynamo
Battery (Volt)	6	6	6	6
Wheelbase (in.)	54·5	54·5	55·5	55·5
Ground clear. (in.)	5·5	5·5	5	5
Seat height (in.)	30		31	31
Width (in.)			26	26
Length (in.)	85		82·5	82·5
Dry weight (lb)	390		380	366
Power: bhp	29	23·5	29·5	25
@ rpm			5500	6000

1 1947—7·12. **2** 1951—3·50.

Model	16H	16H–WD	16H	500T
Year from	**1932**	**1939**	**1945**	**1949**
Year to	**1939**	**1945**	**1954**	**1954**
Bore (mm)	79	79	79	79
Stroke (mm)	100	100	100	100
Capacity (cc)	490	490	490	490
Compression ratio (to 1)	4·9	4·9	4·9	6·0
Valve position	side	side	side	ohv
inlet opens BTDC	0·25 in.	27 to 30°	30	
inlet closes ABDC			73	
exhaust opens BBDC			73	
exhaust closes ATDC	0·25 in.	27 to 30°	30	
Valve clear. (cold) in. (in.)	0·004	0·004	0·002	
Valve clear. (cold) ex. (in.)	0·006	0·006	0·004	
Ignition timing	0·375 in.	0·437 in.	35°	
Points gap (in.)	0·012	0·012	0·012	0·012
Front tyre (in.)	3·25 × 26		3·25 × 19	2·75 × 21
Rear tyre (in.)	3·25 × 26		3·25 × 19	4·00 × 19
Rim front	WM2			WM1
Rim rear	WM2			WM3
Brake front dia. (in.)	7	7	7 **1**	7
Brake front width (in.)	1·25	1·25	1·25	1·25
Brake rear dia. (in.)	7	7	7	7
Brake rear width (in.)	1·25	1·25	1·25	1·25
Front suspension	girder	girder	girder **2**	teles
Rear type	rigid	rigid	rigid	rigid
Petrol tank (Imp. gal)	2·75		2·75 **3**	2·5
Oil tank (Imp. pint)	4		4	3·5
Ignition system	magneto	magneto	magneto	magneto
Generator type	dynamo	dynamo	dynamo	
Output (Watts)			35 **4**	
Battery (Volt)	6	6	6	
Wheelbase (in.)	54·5		54·5	53
Ground clear. (in.)	4·5		5·5	7·2
Seat height (in.)	27·5		29	32
Width (in.)	30		28	28
Length (in.)	84·5		84·5	82
Dry weight (lb)			367	320 **5**
Power: bhp			12·5	21 **6**
@ rpm			4800	5000 **6**

1 1954—8. **2** 1947—teles. **3** 1951—3·5. **4** 1950—60. **5** 1951—300. **6** 1953—24/5300.

Model	Big 4	Big 4–WD	Big 4
Year from	**1932**	**1939**	**1947**
Year to	**1939**	**1945**	**1954**
Bore (mm)	82	82	82
Stroke (mm)	120	120	120 **1**
Capacity (cc)	634	634	634 **1**

Model	Big 4	Big 4–WD	Big 4
Year from	**1932**	**1939**	**1947**
Year to	**1939**	**1945**	**1954**
Compression ratio (to 1)	4·5 **2**	4·8	4·5
Valve position	side	side	side
inlet opens BTDC	0·312 in. **3**	27 to 30	30
inlet closes ABDC			73
exhaust opens BBDC			73
exhaust closes ATDC	0·312 in. **3**	27 to 30	30
Valve clear. (cold) in. (in.)	0·004 **4**	0·004	0·002
Valve clear. (cold) ex. (in.)	0·006 **5**	0·006	0·004
Ignition timing	0·25 in. **3**	0·437 in.	35°
Points gap (in.)	0·012	0·012	0·012
Front tyre (in.)	3·25 × 26		3·25 × 19
Rear tyre (in.)	3·25 × 26		3·25 × 19
Rim front	WM3	WM3	
Rim rear	WM3	WM3	
Brake front dia. (in.)	7	7	7 **6**
Brake front width (in.)	1·25	1·25	1·25
Brake rear dia. (in.)	7	7	7
Brake rear width (in.)	1·25	1·25	1·25
Front suspension	girder	girder	teles
Rear type	rigid	rigid	
Petrol tank (Imp. gal)	2·75		2·75 **7**
Oil tank (Imp. pint)	4		4
Ignition system	magneto	magneto	magneto
Generator type	dynamo	dynamo	dynamo
Output (Watts)			35 **8**
Battery (Volt)	6	6	6
Wheelbase (in.)	54·5		54·5
Ground clear. (in.)	4·5		5·5
Seat height (in.)	27·5		29
Width (in.)	30		28
Length (in.)	84·5		84·5
Dry weight (lb)			373
Power: bhp			14
@ rpm			4500

1 1948—113 mm, 597 cc. **2** 1935—4·8. **3** 1938—0·375 **4** 1938—0·002. **5** 1938—0·004. **6** 1954—8.
7 1951—3·5. **8** 1950—60.

Model	18 & 20	18
Year from	**1932**	**1945**
Year to	**1939**	**1954**
Bore (mm)	79	79
Stroke (mm)	100	100
Capacity (cc)	490	490
Compression ratio (to 1)	6·2	6·45 **1**

Model	**18** & **20**	**18**
Year from	**1932**	**1945**
Year to	**1939**	**1954**
Valve position	ohv	ohv
inlet opens BTDC	0·25 in. **2**	0·312 in.
exhaust closes ATDC	0·25 in. **2**	0·312 in.
Valve clear. (cold) in. (in.)	0·002 **3**	nil
Valve clear. (cold) ex. (in.)	0·004 **3**	nil
Ignition timing (in.)	0·625	0·625
Points gap (in.)	0·012	0·012
Front tyre (in.)	3·25 × 26	3·25 × 19
Rear tyre (in.)	3·25 × 26	3·25 × 19
Brake front dia. (in.)	7	7 **4**
Brake front width (in.)	1·25	1·25
Brake rear dia. (in.)	7	7
Brake rear width (in.)	1·25	1·25
Front suspension	girder	girder **5**
Rear type	rigid	rigid
Petrol tank (Imp. gal)	2·75	2·75 **6**
Oil tank (Imp. pint)	4	4
Ignition system	magneto	magneto
Generator type	dynamo	dynamo
Output (Watts)		35 **7**
Battery (Volt)	6	6
Wheelbase (in.)	54·5	54·5
Ground clear. (in.)	4·5	5·5
Seat height (in.)	27·5	29
Width (in.)	30	28
Length (in.)	84·5	84·5
Dry weight (lb)		374
Power: bhp		21
@ rpm		5000

1 1952—6·6. **2** 1938—0·312. **3** 1938—nil. **4** 1954—8. **5** 1947—teles. **6** 1951—3·5. **7** 1950—60.

Model	**19**	**19**	**19S**	**19R**
Year from	**1932**	**1933**	**1954**	**1954**
Year to	**1932**	**1939**	**1958**	**1955**
Bore (mm)	79	82	82	82
Stroke (mm)	120	113	113	113
Capacity (cc)	588	597	597	597
Compression ratio (to 1)	5·0	5·75	6·2 **1**	6·2
Valve position	ohv	ohv	ohv	ohv
inlet opens BTDC	0·312 in.	0·344 in.	30	30
inlet closes ABDC			75	75
exhaust opens BBDC			78	78
exhaust closes ATDC	0·312 in.	0·344 in.	35	35
Valve clear. (cold) in. (in.)	0·002	0·002 **2**	nil	nil
Valve clear. (cold) ex. (in.)	0·004	0·004 **2**	nil	nil

Model	19	19	19S	19R
Year from	**1932**	**1933**	**1954**	**1954**
Year to	**1932**	**1939**	**1958**	**1955**
Ignition timing (in.)	0·75	0·687	0·687	0·687
Points gap (in.)	0·012	0·012	0·012	0·012
Front tyre (in.)	3·25 × 26	3·25 × 26	3·25 × 19	3·25 × 19
Rear tyre (in.)	3·25 × 26	3·25 × 26	3·25 × 19	4·00 × 18
Rim front			WM2	WM2
Rim rear			WM2	
Brake front dia. (in.)	7	7	8	8
Brake front width (in.)	1·25	1·25	1·25	1·25
Brake rear dia. (in.)	7	7	7	7
Brake rear width (in.)	1·25	1·25	1·25	1·25
Front suspension	girder	girder	teles	teles
Rear type	rigid	rigid	s/a	rigid
Petrol tank (Imp. gal)	2·75	2·75 **6**	2·75 **3**	2·75
Oil tank (Imp. pint)	4	4	4	4
Ignition system	magneto	magneto	magneto	magneto
Generator type	dynamo	dynamo	dynamo	dynamo
Output (Watts)			60	60
Battery (Volt)			6	6
Wheelbase (in.)	54·5	54·5	54·5 **4**	54·5
Ground clear. (in.)	4·5	4·5	6·5	5·5
Seat height (in.)	27·5	27·5	31	29
Width (in.)	30	30		
Length (in.)	84·5	84·5		
Dry weight (lb)			372 **5**	367
Power: bhp			25	25
@ rpm			4200	4200

1 1956—6·4. **2** 1938—nil. **3** 1956—3·25. **4** 1957—57. **5** 1956–393. **6** 1935—3·25.

Model	ES2	ES2	ES2	ES2 Mk 2
Year from	**1932**	**1947**	**1958**	**1964**
Year to	**1939**	**1958**	**1963**	**1966**
Bore (mm)	79	79	79	86
Stroke (mm)	100	100	100	85·5
Capacity (cc)	490	490	490	497
Compression ratio (to 1)	6·2 **1**	6·45 **2**	7·1	7·3
Valve position	ohv	ohv	ohv	ohv
inlet opens BTDC	0·25 in. **3**	30	30	
inlet closes ABDC		75	75	
exhaust opens BBDC		78	78	
exhaust closes ATDC	0·25 in. **3**	35	35	
Valve clear. (cold) in. (in.)	0·002 **4**	nil	nil	
Valve clear. (cold) ex. (in.)	0·004 **4**	nil	nil	
Ignition timing	0·625 in.	0·375 in.	38°	
Points gap (in.)	0·012	0·012	0·015	
Front tyre (in.)	3·25 × 26	3·25 × 19	3·00 × 19	3·25 × 18

Model	ES2	ES2	ES2	ES2 Mk 2
Year from	**1932**	**1947**	**1958**	**1964**
Year to	**1939**	**1958**	**1963**	**1966**
Rear tyre (in.)	3·25 × 26	3·25 × 19	3·50 × 19	3·50 × 18
Rim front		WM2		
Rim rear		WM2		
Brake front dia. (in.)	7	7 **5**	8	8
Brake front width (in.)	1·25	1·25	1·25	1·25
Brake rear dia. (in.)	7	7	7	7
Brake rear width (in.)	1·25	1·25	1·25	1·25
Front suspension	girder	teles	teles	teles
Rear type	rigid **6**	plunger **7**	s/a	s/a
Petrol tank (Imp. gal)	2·75 **14**	2·75 **8**	3·5 **9**	3·5
Oil tank (Imp. pint)	3·5	4	4·5	4·5
Ignition system	magneto	magneto	coil	coil
Generator type	dynamo	dynamo	alternator	alternator
Output (Watts)		35 **10**	60	
Battery (Volt)	6	6	6	6
Wheelbase (in.)	54·75	54·5 **11**	55·5	
Ground clear. (in.)	4·25	6·5	6·2	
Seat height (in.)	27·5	31	31	
Width (in.)	30	28	27	
Length (in.)	84·75	84·5	85	
Dry weight (lb)	396	379 **12**	384	394
Power: bhp		21 **13**		
@ rpm		5000 **13**		

1 1938—6·5. **2** 1952—6·6, 1955—6·8, 1956—7·1. **3** 1938—0·312. **4** 1938—nil. **5** 1954—8. **6** 1939—plunger option. **7** 1953—s/a. **8** 1951—3·5, 1955—2·75, 1956—3·25. **9** 1961—3·62. **10** 1950—60. **11** 1957—57. **12** 1956—389. **13** 1955—25/5300. **14** 1937—3·25.

Model	50	50	50	50 Mk 2
Year from	**1933**	**1955**	**1958**	**1964**
Year to	**1939**	**1958**	**1963**	**1966**
Bore (mm)	71	71	71	72
Stroke (mm)	88	88	88	85·5
Capacity (cc)	348	348	348	348
Compression ratio (to 1)	7·0	7·3	7·3	9·0
Valve position	ohv	ohv	ohv	ohv
inlet opens BTDC	0·344 in.	30	30	
inlet closes ABDC		75	75	
exhaust opens BBDC		78	78	
exhaust closes ATDC	0·406 in.	35	35	
Valve clear. (cold) in. (in.)	0·002 **1**	nil	nil	
Valve clear. (cold) ex. (in.)	0·004 **1**	nil	nil	
Ignition timing	0·687 in.		38°	
Points gap (in.)	0·012	0·012	0·015	
Front tyre (in.)	3·25 × 26	3·25 × 19	3·00 × 19	3·25 × 18
Rear tyre (in.)	3·25 × 26	3·25 × 19	3·50 × 19	3·25 × 18

Model	50	50	50	50 Mk2
Year from	**1933**	**1955**	**1958**	**1964**
Year to	**1939**	**1958**	**1963**	**1966**
Rim front		WM2		
Rim rear		WM2		
Brake front dia. (in.)	7	8	8	8
Brake front width (in.)	1·25	1·25	1·25	1·25
Brake rear dia. (in.)	7	7	7	7
Brake rear width (in.)	1·25	1·25	1·25	1·25
Front suspension	girder	teles	teles	teles
Rear type	rigid	s/a	s/a	s/a
Petrol tank (Imp. gal)	2·75	3·25	3·5 **2**	4
Oil tânk (Imp. pint)	3·5	4	4·5	4
Ignition system	magneto	magneto	coil	coil
Generator type	dynamo	dynamo	alternator	alternator
Output (Watts)		60	60	
Battery (Volt)	6	6	6	6
Wheelbase (in.)	54	54·5 **3**	55·5	
Ground clear. (in.)	4·5	6·5	6·2	
Seat height (in.)	27	31	31	
Width (in.)	30	27	27	
Length (in.)	82	85	85	
Dry weight (lb)		382	377	382
Power: bhp		20		
@ rpm		5800		

1 1938—nil. **2** 1961—3·62. **3** 1957—57.

Gearbox and transmission data

To avoid considerable repetition this has been omitted from the specification tables of the road models and is given next. To establish the box and overall ratios for any year it is necessary to determine which box was fitted and the top gear for the machine. Simple multiplication will give the indirects.

The period from 1932 saw changes in the gearbox which stabilised in 1935 with the adoption of the Norton unit. However, different internal ratios could be fitted to suit the user and the overall gear was often altered by changing the engine sprocket. The most common reason for this was the fitting of a sidecar when the normal practice was to reduce the engine sprocket by two teeth.

Where the racing bottom gear (1·77) was fitted the small diameter of the layshaft gear precluded the fitting of a kickstarter. This applied to the Manx and those camshaft models supplied for racing.

In the Clubmans, and at Daytona, a kickstarter was compulsory so a lower bottom gear ratio was fitted which allowed this facility to be provided. This gave a big jump from second to first and due to this some riders came off at Daytona, no doubt caught out as they changed down for the hairpin bend after two miles of straight. They found it better to avoid bottom gear except when moving off.

Gearbox ratios

1932 **3-speed**—standard	1, 1·27, 2·4
special close	1, 1·33, 1·77
wide	1, 1·47, 2·94
4-speed—standard	1, 1·21, 1·61, 2·98
International—with kickstart	1, 1·1, 1·33, 2·33
International—no kickstart	1, 1·1, 1·33, 1·77
International—Clubmans/Daytona	1, 1·1, 1·33, 2·12
1935 **Norton** box	1, 1·21, 1·77, 2·97
Big 4–WD and **1949 500T**	1, 1·47, 2·39, 3·67
1950 500T	1, 1·47, 2·39, 3·31
AMC box	1, 1·35, 1·77, 2·67
AMC box from 1960	1, 1·22, 1·7, 2·55

Transmission

Clutch has 42 teeth
Gearbox has 19 teeth except for **500T** with 16 and late model **50** when fitted with sidecar when 17 were used.

Rear wheel: up to 1952	42 teeth
1953 on	43 teeth
except **16H–WD** which had	43 teeth
and **Big 4–WD** which had	52 teeth

Engine sprockets

At least three varieties of sprocket offset exist as this feature was changed in 1948 when the timing gear was altered and again when the alternator was adopted.

Year	Models	Engine sprocket	Overall gearing	Remarks
1932–37	16H, Big 4, 30	20	4·64	
	18, 19, 20, ES2, CS1	21	4·42	
	40, 50, 55	18	5·16	
	CJ	17	5·46	
1938–39	16H, Big 4	19	4·89	
	18, 19, 20, ES2, CS1, 30	20	4·64	
	50, 55, CJ, 40, 16H–WD	18	5·16	
1940–45	16H–WD	18	5·28	rear 43T
	Big 4–WD s/c wheel drive	18	6·39	rear 52T
	16H–WD civilian use	19	5·00	rear 43T
	Big 4–WD civilian use	21 or 22	5·47 or	
			5·22	rear 52T
1946–58	30	20	4·64	1946—std box
	40	18	5·16	1947 on—close
1949–54	500T	20	5·51	gearbox 16T
1945–52	16H	19	4·89	
1953–54	16H	19	5·00	rear 43T
1945–52	18 and ES2	20	4·64	
1953–63	18 and ES2	20	4·75	rear 43T
1947–52	Big 4	17	5·46	
1953–54	Big 4	17	5·60	rear 43T
1954–58	19S	21	4·52	rear 43T
1954–55	19R	18	5·28	rear 43T
1955–63	50	18	5·28	rear 43T
1964–66	5 Mk 2		5·51	
1964–66	ES2 Mk 2		4·79	

Chains

All primary chains were $\frac{1}{2} \times \frac{5}{16}$ and all final drive $\frac{5}{8} \times \frac{1}{4}$ for all real Nortons. The AMC inspired Mk 2 models of 1964–66 used the same primary but had $\frac{5}{8} \times \frac{3}{8}$ final drive chain. Racing models used racing quality chain which could be fitted to any machine.

Colours

During the 1930s machines were finished in black with only the petrol tank, oil tank and sometimes the wheel rims to relieve this, along with the exhaust system, headlamp rim, controls and minor items which were chrome plated.

Tank finish was silver or dull nickel plated for the racing machines in the late 1930s with side panels lined with outer $1\frac{3}{8}$ in. black line and thin red inner line. This was continued, with black or alloy rims, up to 1951 when the featherbed appeared still with two side panels lined black and red but in silver only and with the black reduced to $\frac{7}{8}$ in.

The International followed the above but with a $1\frac{1}{4}$ in. black line. Late 1930s and all postwar models up to 1953, and the adoption of the featherbed, had chrome plated tanks, the petrol with silver panels top and sides with $\frac{5}{8}$ in. black outer and thin red inner lines. The oil tank was lined but without silver panels and the black width remained at $1\frac{1}{4}$ in.

Standard models in the 1930s had silver tanks panelled with $\frac{1}{2}$ in. black outer and thin red inner lines, also black rims and chrome fittings. Some had chrome plated tanks and then the petrol had silver panels but not the oil which was lined on the chrome, all lining remaining at $\frac{1}{2}$ in. black and thin red. With this finish the chrome fittings would continue plus chrome wheels with black centres lined red.

Standard ohc and other models from 1935–38 had the petrol tank lining reversed so the thin red line was outside the black one. The oil tanks were not necessarily the same.

1945

16H and **18**: black with chrome plated petrol and oil tanks with silver panels lined with $\frac{1}{2}$ in. wide black outer and thin red inner line. Wheel rims chrome plated with black centres lined red. Dome on chaincase over engine sprocket chrome plated.

1946

16H and **18**: no change.

1947

16H and **18**: chrome plated headlamp rim.

ES2 and **Big 4**: as **16H** and **18**.

Inter: black with silver tanks lined $\frac{5}{8}$ in. wide petrol and $1\frac{1}{4}$ in. oil in black with thin red inner, wheels as model **18**.

Manx: as Inter with black lining $1\frac{3}{8}$ in. wide, wheel rims black.

1948–49–50

as 1947 except no chrome dome on chainguard, which became all black.

500T: black frame, fork shrouds, mudguards, oil tank, toolbox; petrol tank dull chrome plated with silver top panel with $\frac{1}{2}$ in. wide black and thin red lines.

1951

as 1947 but during year, due to a nickel shortage, tank finish altered to all silver with panels lined in red and black as before. Wheel rims continued as before but some finished all silver with red lining.

1952

new tank panel style with thinner ($\frac{1}{4}$ in.) black line for road models (not ohc), chrome rims with silver centres lined red, black oil tank.

1953

Manx: black with silver tanks with $\frac{7}{8}$ in. black and thin red lines, alloy rims.

Inter: polychromatic grey frame, forks, headlamp shell, oil tank, mudguards, chainguard, hubs.

Rims chrome plated with silver centres lined grey. Petrol tank chrome plated with polychromatic grey panels with outer black and inner red lines. Norton transfer in black with white outline on tank sides, kneegrips fitted. Option of **Manx** black and silver available

ES2 and **18**: black with petrol tank chrome plated with silver panels with black outer and red inner lines. Oil tank black, wheel rims chrome plated with silver centres lined red.

16H and **Big 4**: as **ES2** except tank silver with black and red lining as in late 1951.

1954

Prototype 250: silver grey, wheel rims chrome plated with silver centres lined red.

other models: as 1953.

1955

Manx: as 1953.

Inter: as 1953 with round plastic tank badges as **ES2**.

ES2, 19S, 19R: as **ES2** in 1953 with round plastic tank badge. Each badge held by two screws with red Norton insignia on a silver letter 'N' which was relieved with horizontal lines. The letter was superimposed on a round quartered background. Top right and lower left quadrants were black, the others clear with vertical lines. The quadrants were encircled by rings in silver, clear and black. Machine fork sliders were polished and the more prominent nuts and bolt heads polished after plating.

1956

all models: as 1955.

50: as **ES2** and **19S**.

1957

Manx and **Inter**: as 1955.

All ohv: as 1956 except tank painted matt silver with

separate chrome plated panels on plastic edge bead. Each panel carried round badge and kneegrip.

1958

all models: as 1957.

1959

Manx: as 1953.

50 and **ES2**: forest green all painted parts with chrome plated tank side panels as in 1957, optional finish in black and silver as 1957.

1960

Manx: as 1953.

50 and **ES2**: as 1959 with option of chrome plated mudguards also available.

1961

Manx: as 1953.

50 and **ES2**: in two-tone with new long tank badge styled to give dividing line to tank. Badge based on long rectangular shape with Norton name at front, section with horizontal lines in centre, and kneegrips at rear all underlined by shaped line. White dividing line, $\frac{1}{8}$ in. wide, on front and rear of tank sides. Petrol tank top, fork shroud, headlamp, frame, oil tank, battery cover, chaincase, chainguard in black for model **50** and green for **ES2**. Lower tank and mudguards dove grey for both. Option of chrome plated mudguards.

1962

Manx: final model on same lines as 1953 and before with black frame, top fork shroud, front number plate and fairing, mudguards, chainguard; petrol tank in silver with broad black outer panel lining and thin red inner; silver oil tank; alloy rims; chrome plated exhaust system, fork seal cups, minor controls; seat in black with red piping.

50 and **ES2**: as 1961, chrome plated mudguard option continued.

1963

50 and **ES2**: as 1961 in black and off-white for both models, the latter replacing the dove grey for the lower tank and mudguards.

1965

50 Mk 2 and **ES2 Mk 2**: black frame, forks, headlamp shell, mudguards, oiltank, battery cover; silver petrol tank with round plastic badges and tank lining.

1966

50 Mk 2 and **ES2 Mk 2**: as 1965 with altered tank lining and Norton transfers on side panels.

Engine and frame numbers

The Norton numbering system in the postwar years was based on the use of a letter to indiciate model year, a number to indicate the model and a further number for the machine, the same one being used for engine and frame. This last ran in sequence from 1001 on without regard to the model and for a long time without any gaps. Only in much later years do gaps occur in the numbers where batches were allocated (one assumes) but not built.

The engine number normally also had the engine dimensions stamped by it on the left side of the crankcase just below the barrel. In 1947 some ohv engines were fitted with cams with quietening ramps and these had the letter Q added after the engine number.

The year letters used were:

A	1946	F	1951	L	1956
B	1947	G	1952	M	1957
C	1948	H	1953	N	1958
D	1949	J	1954	P	1959
E	1950	K	1955	R	1960

after which the practice was discontinued due to the large stocks of 1960 machines left unsold at the end of the year.

These letters were applied to the model year which usually began in the autumn so that new models could be prepared in time for the Earls Court Show and stocks delivered to dealers.

Thus the letter generally applies to machines built from about September in one year to August in the next. This is not 100 per cent true as there can be some overlap as model changes were taken up in production so that, for example, H 48000 could be followed by G 48001 if the latter was to the older specification or had been held up in production.

To assist dating, the first numbers of the New Year are listed below. These are taken from the Norton books but again have to be treated as approximate as after the 1940s there is no definite change point, the book dates jumping back and forth for a number of machines.

1945	1001	1952	42700	1959	80488
1946	2131	1953	48900	1960	87038
1947	7756	1954	55350	1961	94500
1948	13792	1955	60700	1962	101060
1949	20701	1956	66600	1963	105000
1950	27100	1957	71360	1964	108138
1951	35560	1958	77400	1965	111650

The model numbers used were:

Model 2	16H	9	19S
3	18	10	International 40
3T	500T	10M	Manx 40
4	ES2	10M2	Manx 40 featherbed
6	19R	11	International 30
7	Big 4	11M	Manx 30
8	Spring frame 16H or Big 4	11M2	Manx 30 featherbed
		13	50

Frame numbers are stamped in the headstock region on the tank support lug left side or, on featherbed frames, on the left rear frame gusset above the fork pivot. In this case the number itself is in a vertical line with the model number horizontal.

Carburettor settings

Model	Year	Amal type	Size	Main	Pilot	Slide	Needle pos.	Needle jet
40M	1946–51	RN	$1\frac{3}{32}$	500		6	5	·109
40M	1952–54	GP	$1\frac{1}{8}$	230		5	3	·109
40M	1955–57	GP	$1\frac{5}{32}$	270		5	3	·109
40M	1959–60	GP	$1\frac{7}{32}$	280		4	3	·109
40M	1962–63	GP	$1\frac{5}{16}$	390	25	5	3	·109
30M	1946–51	RN	$1\frac{3}{16}$	560		6	4	·109
30M	1952	GP	$1\frac{7}{32}$	260		5	3	·109
30M	1953–56	GP	$1\frac{9}{32}$	370		5	3	·109
30M	1956–60	GP	$1\frac{5}{16}$	390		5	3	·109
30M	1962–63	GP	$1\frac{3}{8}$	390	25	5	3	·109
40	1933–34	TT	$1\frac{1}{16}$	51		4	4	·1075
40	1935–37	TT	$1\frac{1}{16}$	260		4	4	·1075
40	1937–39	TT	$1\frac{3}{32}$	350		6	4	·109
40	1938	TT	$1\frac{3}{32}$	460		6	1	·109
40	1939	TT	$1\frac{3}{32}$	500		6	5	·109
40	1946–54	TT	$1\frac{3}{32}$	350		6	4	·109
40	1956–58	GP	$1\frac{3}{16}$	290		5	3	·109
30	1933–34	TT	$1\frac{1}{8}$	56		5	3	·109
30	1935	TT	$1\frac{1}{8}$	310		5	3	·109
30	1937–39	TT	$1\frac{5}{32}$	460		7	2	·109
30	1937	TT	$1\frac{5}{32}$	310		5	3	·109
30	1938	TT	$1\frac{5}{32}$	560		6	3	·109
30	1939	RN	$1\frac{5}{32}$	600		6	4	·109
30	1946–54	TT	$1\frac{5}{32}$	350		7	2	·109
CS1	1933–37	89	$1\frac{3}{32}$	180		29/4	3	
16H	1933–37	76	1	160		6/4	3	
16H	1938	76	1	170		6/4	3	
16H	1939	76	1	160		6/5	3	
16H-WD	1939	276	1	170		6/4	3	
16H-WD	1940–45	276	1	170		6/4	2	
16H	1946–54	276	1	160		6/5	3	
18	1937–39	76	$1\frac{1}{16}$	160		6/4	3	
18	1946–54	276	$1\frac{1}{16}$	160		6/4	3	

19	1933–39	76	$1\frac{1}{16}$	160		6/4	3	
19	1955–58	376	$1\frac{1}{16}$	270	30	4	3	·106
20	1933–39	76	$1\frac{1}{16}$	200		6/4	3	
ES2	1937–39	76	$1\frac{1}{16}$	160		6/4	3	
ES2	1946–54	276	$1\frac{1}{16}$	160		6/4	3	
ES2	1955–63	376	$1\frac{1}{16}$	270	30	4	3	·106
ES2 Mk 2	1966	389	$1\frac{1}{8}$	290	25	$3\frac{1}{2}$	3	·106
Big 4	1933–35	76	1	160		6/4	3	
Big 4	1937–38	76	1	170		6/4	3	
Big 4	1939	76	1	160		6/5	3	
Big 4	1946–49	276	1	160		6/5	3	
Big 4	1950–54	276	1	170		6/4	3	
50	1933–39	76	1	170		6/4	3	
50	1956–63	376	1	210	30	$3\frac{1}{2}$	2	·106
50 Mk 2	1966	389	$1\frac{1}{8}$	260	25	3	3	·106
Trials 350	1946–49	276	1	170		6/4	3	
500 T	1948–54	276	$1\frac{1}{16}$	160		6/4	3	

Prices

The UK prices of the machines during the postwar period are set out in the tables below. It is worth noting that many of the variations during the early 1950s were due to changes in the purchase tax imposed in those days and that the basic machine price would alter little or nothing over several years. A change from 20 years later.

Date	Manx 30M & 40M	Inter 30	Inter 40	350 & 500 trials	500T
26. 9.46	£298. 9*s*. 0*d*.	£229.17*s*. 5*d*	£220.19*s*. 7*d*	£171. 9*s*. 0*d*.	
8. 5.47	£314.19*s*. 2*d*.	£247. 0*s*. 4*d*.	£238. 2*s*. 6*d*.	£183.10*s*. 4*d*.	
11.11.48	£314.19*s*. 2*d*.	£247. 0*s*. 4*d*.	£238. 2*s*. 6*d*.		£177. 3*s*. 4*d*.
13.10.49	£346.14*s*. 2*d*.	£252. 1*s*.11*d*.	£243. 4*s*. 1*d*.		£182. 5*s*. 0*d*.
8. 3.51	£346.14*s*. 2*d*.	£264.15*s*.11*d*.	£255. 5*s*. 5*d*.		£198. 2*s*. 5*d*.
5. 7.51	£426.14*s*. 5*d*.	£278. 2*s*. 7*d*.	£267.19*s*. 5*d*.		£208. 5*s*. 7*d*.
6. 9.51	£429. 6*s*. 8*d*.	£279.16*s*. 8*d*.	£269.12*s*. 3*d*.		£209.11*s*. 1*d*.
2.10.52	£429. 6*s*. 8*d*.	£279.16*s*. 8*d*.	£269.12*s*. 3*d*.		£199. 6*s*. 8*d*.
23. 4.53	£406. 0*s*. 0*d*.	£264.12*s*. 6*d*.	£254.19*s*. 2*d*.		£188.10*s*. 0*d*.
17. 9.53	£418.16*s*. 0*d*.	£273.12*s*. 0*d*.	£266. 8*s*. 0*d*.		£194. 8*s*. 0*d*.
16. 9.54	£465.12*s*. 0*d*.	£273.12*s*. 0*d*.	£266. 8*s*. 0*d*.		
3.11.55	£481. 2*s*. 5*d*.	£282.14*s*. 5*d*.	£275. 5*s*. 7*d*.		
5. 9.57	£496.10*s*. 1*d*.	£303. 2*s*.10*d*.	£295.13*s*. 2d.		
16. 4.59	£480. 1*s*. 9d.				
11. 2.60	£496.19*s*. 6*d*.				
8. 3.62	£536.16*s*. 0*d*.				
12. 4.62	£528. 0*s*. 0*d*.				

Date	16H	Big 4	18	ES2	19R	19S	50
20. 9.45	£124. 8s. 2d.		£131.19s. 0d.				
7. 3.46	£140. 6s. 8d.		£147. 6s. 5d.				
15. 8.46	£149. 4s. 6d.		£156. 4s. 3d.				
26. 9.46	£158.15s. 0d.	£162.17s. 6d.	£165. 2s. 0d.	£180. 6s. 9d.			
8. 5.47	£170.16s. 4d.	£175. 5s. 2d.	£177. 3s. 4d.	£193.13s. 6d.			
13.10.49	£175.17s.11d.	£180. 6s.10d.	£182. 4s.11d.	£198.15s. 1d.			
8. 3.51	£184.15.s. 8d.	£189. 4s. 7d.	£191. 2s. 8d.	£208.18s. 4d.			
5. 7.51	£194.16s. 2d.	£198.15s. 1d.	£200.13s. 2d.	£219. 1s. 6d.			
6. 9.51	£195.10s. 0d.	£199.19s. 5d.	£201.17s. 9d.	£220. 8s. 4d.			
2.10.52	£186.11s. 1d.	£190. 7s. 9d.	£191.13s. 4d.	£209.11s. 1d.			
23. 4.53	£176. 8s. 4d.	£180. 0s.10d.	£181. 5s. 0d.	£198. 3s. 4d.			
17. 9.53	£182. 8s. 0d.	£186. 0s. 0d.	£187. 4s. 0d.	£204.12s. 0d.			
16. 9.54				£204.12s. 0d.	£190.16s. 0d.	£208. 4s. 0d.	
3. 2.55				£204.12s. 0d.	£198.12s. 0d.	£216. 0s. 0d.	
15. 9.55				£212. 8s. 0d.		£222. 0s. 0d.	£206. 8s. 0d.
1.12.55				£219. 9s. 7d.		£229. 8s. 0d.	£213. 5s. 7d.
5. 9.57				£242.12s. 9d.		£254. 9s.10d.	£236. 8s. 0d.
16. 4.59				£234.12s. 4d.			£228.11s. 8d.
5. 5.60				£246. 1s. 6d.			£240. 0s.10d.
21. 9.61				£250. 2s. 0d.			£244. 0s. 0d.
12. 4.62				£258. 0s. 0d.			£252. 0s. 0d.
				ES2 Mk 2			**50 Mk 2**
19.11.64				£282.19s. 0d.			£279.15s.11d.
9. 9.65				£297.16s. 5d.			£290. 1s. 3d.

Norton model numbers

The apparently unrelated Norton model numbers do run in some sequence and are listed below. I am indebted to an article Titch Allen wrote for *Motorcycle Sport* for this information and he in turn acknowledges that the original work was done by Geoffrey Wilson.

All Norton single enthusiasts owe them a debt

1 Big 4 633 cc, 3-speed Sturmey Archer gearbox
2 490 cc, 3 speed Sturmey Archer gearbox
3 Big 4 633 cc, 3-speed Sturmey box, belt drive
4 490 cc 3-speed Sturmey box, belt drive
5 Big 4 633 cc, 3-speed hub, belt drive
6 490 cc, 3-speed hub, belt drive
7 Brooklands Special, 490 cc, direct belt drive
8 Brooklands Road Special, 490 cc, direct with Phillipson pulley
9 Tourist Trophy 490 cc, direct belt or Phillipson pulley
10 490 cc, Touring machine, direct belt drive
11 Big 4 633 cc, direct drive touring machine
12 490 cc, Colonial model, 3 speed hub, belt drive
13 Big 4 633 cc, Colonial, 3-speed hub, belt drive
14 Big 4 633 cc, 4-speed, all chain
15 Big 4 633 cc, 3-speed, De Luxe Sidecar outfit
16H 490 cc, side valve, 3-speed, chain drive touring model
17 490 cc, side valve, 3-speed, chain drive colonial model
18 490 cc, ohv, 3-speed, chain drive
19 588 cc, ohv, 3-speed, chain drive
20 490 cc, ohv, twin port
21 490 cc, ohv, as model 18 but dry sump
22 490 cc, ohv, twin port, cradle frame, 3-speed
24 588 cc, ohv, 4-speed
25 490 cc, ohv, TT replica, 3-speed, no kickstart
ES2 490 cc, ohv, 3-speed, cradle frame
JE 348 cc ohv, 3-speed, cradle frame
CS1 490 cc, ohc, 3-speed, cradle frame

CJ 348 cc, ohc, 3-speed, cradle frame
30 490 cc, ohc, 4-speed, International
34 490 cc, ohv, 4-speed, 1927
40 348 cc, ohc, 4-speed, International
44 588 cc, ohv, 4-speed, 1927
50 348 cc, ohv, 4-speed, 1933
55 348 cc, ohv, 4-speed, twin port

There are various stories as to how the ES2 acquired its designation. The first is that it was a tribute to Ernest Searle, which is romantic but very unlikely. The second is that it stood for 'Enclosed Springs' and model '2' of the 18 line, and it is this story that Edgar Franks agrees to be the correct one. However, Walter Moore supported theory number three that all the letters indicated some aspect of the machine thus:

C Camshaft
E Extra cost
J Junior, i.e. 350 cc
S Sports

and hence: ES2 meant Extra cost and Sports with '2' indicating the second of the sports line with 18 being the first.
JE meant Junior and the more Expensive cradle frame.
CJ meant Camshaft with 350 cc engine.
CS1 meant first Camshaft Sports model.

Franks' opinion seems the most likely as in his role in the drawing office he would have been more involved with such matters than Moore who was concerned more with design and engineering rather than the details of model designation, which is basically an administration point.

This seems very likely to me from my own industrial experience but I would suggest that, as the CS1 and ES2 both appeared in 1928, maybe they were considered as models 1 and 2 of the cradle frame series with the letter 'C' indicating the ohc and that later the letters were used in the way suggested by Moore to indicate some feature.

I do know that often such matters are decided by either the drawing office or designer in an arbitrary way and even if some system is intended to be followed it often falls by the wayside very quickly.

Works racers

This appendix is a précis of the main chapters and gives details of the works racing machines over the years so that the development of the main team machines and the appearances of the experimental ones can be quickly determined.

1922
Prototype ohv built and run in TT where it retired.

1923
Ohv machine run in TT.

1924
TT wins by ohv, year of firsts.

1927
Moore ohc engine first used.

1929
Moore moved to NSU.

1930
Carroll engine first used, horizontal inlet tract, oil drain from cambox, 3-stay frame, 3-speed non-positive gearbox. For Ulster, engine had downdraught inlet tract, 2-stay frame, rear brake on right.

1931
No oil drain pipe from cambox, 4-speed gearbox, 2-stay frame, rear brake on left.

1932
More cooling fins, many parts in light alloy, fork check springs of large parallel form type, narrow front hub in one piece with drum.

1933
Aluminium bronze alloy head, iron barrel.

1934
Aluminium fins cast to bronze head skull, aluminium barrel with iron liner, hairpin valve springs, bolt through tank, twin plugs, megaphone.

1936

Plunger frame, front fork spring on spherical seating, 597 cc sohc engine built for sidecar use.

1937

Twin overhead camshafts, more extensive fins, single plug, rev-counter, alloy rear sprocket, coil valve springs fully enclosed tried.

1938

Revised engine dimensions, square head, new crankshaft with vertical fins, revised bevel box, barrel sunk more deeply into case, new frame under tubes, much neater plungers, telescopic front forks, conical hub, ribbed drums, air scoop.

1947

Roadholder front forks, Newton rear dampers.

1948

Common crankcase for both engine sizes with four studs at base, hairpin spring legs in slots, revised rear engine mounting, magneto mounted higher, new float type, smaller pipe on 350, oil tank breather, shorter front forks, steering damper below bottom crown, Newton rear dampers, 19 in. rear wheel, twin leading-shoe brakes front and rear, 4-plate clutch, primary chain covered in by outer plate, guard for top run of final chain, rev-counter mounted on right fork leg.

1949

New forks with spindle ahead of sliders, steering damper spring added, new front hub with large hollow spindle, separate air vent on petrol tank, oil tank cap at angle, standard float on left mounted to seat tube, special Amal tried on some engines, 597 cc dohc engine for Eric Oliver.

1950

Featherbed frame, bolted sub-frame, fork pivoted on Silentblocs, rear units with remote reservoirs, Roadholders, clip-ons, welded-on control lever pivots, single seat with cut out for oil filler, small tail fairing carrying rear racing numbers, central oil tank with float chamber platform on front face, petrol tank held by strap onto rubber mountings, 19-in. wheels with 3·00 front and 3·25 in. rear tyres, exposed primary chain, laid down gearbox with remote change linkage, shorter exhaust, smaller megaphone, GP carburettor, engine dimensions 73·32 × 82.5 mm and 82 × 94·3 mm, new cylinder head with cast-in bevel box, iron valve inserts, closer pitched fins, diagonal fins across top of head, modified cambox with longer bosses on left and rev-counter drive, filter in lower bevel chamber, engine steady.

1951

Engine dimensions 75·9 × 77 mm and 84 × 90 mm, cambox bolted straight to head, GP carburettor, suspended float chamber, steeper inlet tract and reverse cone megaphone on 350, swept back exhaust pipe, clutch cable entry to gearbox altered, smaller clutch worm with thrust ball race, lighter gauge frame tubes, no remote reservoirs on rear units, internal springs for forks, no rear tail fairing, rubber mounted mudguards, 3·50 × 18 in. rear tyre, remote rear brake not used.

1952

500 engine dimensions 85·93 × 86 mm, fins extended round vertical drive shaft tube which was in light alloy, modified carburettor bellmouth, rotating magnet magneto, welded sub-frame and seat support, rear units inclined and with cast alloy spring boxes, sintered bronze rear fork bushes, rear brake on right remote from sprocket and cable operated, rear tyre on 500 4·00 × 16 in. on WM3 rim. Rotary valve engine built.

1953

500 engine dimensions 88 × 82 mm, oil cooled exhaust valve guide, finned lower bevel housing, weir float, fuel pump driven by inlet camshaft, experimental gearbox, clutch and engine shock absorber, twin outlets for primary chain oiler, guard extended round clutch, rear chain oil in left sub-frame tube, oil tank matt black, petrol tank recesses modified, 3·00 × 19 front and 3·50 × 19 in. rear tyres used. Kneeler model tried using 350 cc engine in special low frame with very extensive streamlining and pannier fuel tanks. Trailing link fork tried.

1954

Engine dimensions 78 × 73 mm and 90 × 78·4 mm, shorter rod, stiffer crankshaft, outside flywheel, engine shaft shock absorber, flap valve breather in front of crankcase, shorter exhaust pipes, Burman 5-speed gearbox with direct pedal, 5-plate clutch, primary chain oil in top left frame tube, weir carburettor without pump return, AC pump used with pannier tanks, frame altered to clear flywheel, sub-frame modified, 350 made from tubing one gauge thinner, brake cross shaft through rear fork pivot, extractor fan front hub, proboscis fairing, no rear damper shrouds.

1955

Works used standard Manx modified with weir float, rotating magnet magneto, magnetic rev-counter, exhaust pipe and megaphone in one, modified cams, forks with external springs, phosphor bronze rear fork bushes, Woodhead Monroe rear units with exposed springs, rubber covered tubular footrests, a bracing strut on the gear pedal, Renold chain oiler, primary chain oil in top left frame tube, primary chainguard extended, rear chain buffer, smaller front mudguard without stays, stiffer rear mudguard, lower and softer seat, welded-on lever pivots with spring locked adjustors, smaller flyscreen, no air lever. F model—flat engine in 350 and 500 cc sizes to general works specification with two-piece crankshaft, integral 5-speed gearbox, two layshafts, cross-over drive, dry clutch, spine frame, telescopic front forks, swinging fork rear, extractor front hub, 8 in. remote rear brake with 2 ls, oil carried in frame.

1956

Carburettor chokes up $\frac{1}{32}$ in. for each size engine, rev-counter rubber mounted into top of number plate fairing itself rubber mounted to the forks, machine height lowered with shorter fork stanchions, modified sub-frame with no rear loop behind seat, ribbed front brake drum.

1957

Proboscis fairing and pannier tanks used, also 90 bore engine with internal flywheels only and a longer rod, desmodromic valve operation tried on bench.

1958

Work with desmo engine led to redesign of cam drive. Junior 80 bore engine with dimensions of 80 × 69·5 mm and coil valve springs tried in Duke frame with leading link front forks. 93 bore engine built using eccentric sleeve on crankpin, 73·5 mm stroke. Big bearing crank engine built using 5·5 in. bore drive side main and outside flywheel with cover plate.

1959

Desmo 500 and 350 with dimensions of 86 × 60 mm tried at TT.

1960

Desmo engine with dimensions of 86 × 60 mm and outside flywheel built. Lowboy built with special bicycle type front forks and fitted with standard engine, later fitted with chest support for rider.

1961

7 in. dual front brake, later in years a shortened 350 engine was tried in the Domiracer frame.

Model recognition

These notes are to some extent a précis of the main text and may need to be used in conjunction with it and the other appendices.

1931

CS1: Carroll engine, left pipe, 3-speed handchange, 3-stay frame, step in front engine plates to suit smaller crankcase.
CJ: Carroll engine, right pipe, 3-speed handchange 2-stay frame, step in front engine plates to suit smaller crankcase.
Big 4, **16H**, **18**, **19**, **20**, **22**, **ES2**: new style engine with magneto behind cylinder driven from inlet camshaft, dry sump lubrication, new oil pump, oil tell-tale in timing case, pushrods in tubes, detachable head for side valves, left pipes.
ES2: in **CS1** cradle frame, rest in diamond 2-rear stay type. All with Norton forks, not Norton hubs, 3-speeds.

1932

Inter: introduced in 350 and 500 sizes, 4-speed footchange.
All ohc: right pipes.
19: through studs for head and barrel.
All models: Norton qd wheels, interchangeable.
Sv: right pipes, steel toolbox.

1933

Inter: Norton forks with check springs.
CS1, **CJ**, **ohv**: 4-speeds as standard, positive stop option.
19: dimension change to 82 × 113 mm.

Sv: 3-speeds standard, 4 option.
Introduction of **50** and **55** models of 348 cc, deletion of **22**.
50 and **55**: through studs for head and barrel.

1934
Ohc: oil supply increased, filter bolt added to bottom bevel box.
Ohv and **sv**: second main on drive side, timing gears as train, valve guide oil pipes added to **ohv**, tubular silencers.
All: no engine shock absorber, Norton clutch, oil bath primary case, Norton forks with check springs, neater positive stop mechanism.
Inter: one-piece front hub and drum and narrower front fork.

1935
All: Norton 4-speed footchange gearbox.
Inter: aluminium bronze head option, hairpin valve springs.
Ohv: right pipes.

1936
Inter: 14 mm plug, light alloy/bronze skull head, light alloy/iron liner barrel.
Ohv: larger diameter pushrod tubes, open wheel hubs.
Sv: open wheel hubs.
18, **20** and **ES2**: through studs for head and barrel.

1937
Ohv: 14 mm plugs.
All road: Lucas CVC dynamo control system.

1938
Inter: option of plunger frame.
Ohv: oil tell-tale deleted, enclosed valve gear, pushrod tubes angled in at top, twin pipe silencer.
Sv: tell-tale and silencer as **ohv**, enclosed valves.

1939
Manx Grand Prix introduced—**Inter** in plunger frame to racing specification, megaphone exhaust, conical hubs, bolt-through tank, wrap-round oil tank, single ohc, square head fins.
CS1, **CJ**: plunger option.
ES2: plunger option, **Inter** type tank option.
All: tubular silencer.

1945
16H, **18**: as 1939 in cradle frame with girders, enclosed clutch worm end cover, larger toolbox, altered oiltank, tubular silencer.

1946
16H, **18**: no change.
Manx: alloy engine, square head, hairpins, sohc, teles, plungers megaphone, saddle and pad, quick action caps.

1947
Manx: as 1946.
Inter: iron engine, sohc, teles, plungers, enclosed clutch mechanism on gearbox with circular access cap, saddle.
16H, **18**: teles with speedometer in fork bridge.
ES2: as **18** in frame with plungers.
Big 4: as **16H** with 633 sv engine with longer barrel and tappets as on WD model.
During year clutch lever access cap changed to cover secured by two screws.
Silencers were tubular, offset inlet and exit for **ohv**, in-line for **sv** and **ohc**.
Ohv: some fitted with Q cams.

1948
Manx, **Inter**: no change.
Sv, **ohv**: smaller timing case, no cam followers, flat non-rotating tappets, dished cam gears, smaller but wider flywheels, longer piston, oil pump gear with pinholes, magneto sprocket teeth number altered.
Sv: alloy casting enclosing valve gear, alloy head, 14 mm plug.
Ohv: one-piece rocker box with fixed spindles, oil supply from main return pipe, in-line silencer.
Big 4: stroke reduced to 113 mm and capacity to 597 cc.

1949
Manx: at TT models with dohc and alloy tanks.
Others: brake shoe pivot tie plate added, central rear lamp, **ohv only** alloy pushrods.
500T: alloy engine, special fork yokes.

1950
Manx, Inter: no change.
Sv, **ohv**: laid down gearbox, common juction for oil pipes, CVC in toolbox.
ES2: prop stand, extended chainguard.
500T: common oil pipe junction, altered first gear ratio.

1951

Manx: featherbed frame, 1950 motor, gauze filter in lower bevel chamber, cambox rev-counter drive, laid down gearbox, 19 in. wheels, internal fork springs.
Inter: no change.
Sv, ohv: 3·5 gallon petrol tank, new form 4-pint oil tank, die-cast alloy front brake backplate, prop stands on all.
500T: no front hub scollops.

1952

All models: no change.

1953

Manx: no change.
Inter: featherbed frame, alloy head and barrel, down-draught TT, pear shaped silencer, laid down gearbox, 8 in. front brake, underslung pilot, Diacon rear light.
ES2: swinging fork frame, Girling rear units, smaller exhaust pipe, pear shape silencer, qd rear wheel as rigid models, 43T rear wheel sprocket, underslung pilot, Diacon rear light, dualseat.
16H, **18**, **Big 4**: dualseat, Diacon rear light, 43T rear wheel sprocket.
500T: die-cast head.

1954

Manx: short stroke engines, integral top bevel housing, barrel fins round vertical drive tube, diagonal fins between valves, cambox bolted direct to head, steeper inlet tract, welded sub-frame, 8 in. 2ls front brake, works shape to oil tank.
Inter: no change.
Sv, **ohv**: 8 in. front brake.
500T: rumoured Lucas magneto, kept 7 in. brake.
Late in year **16H**, **Big 4**, **18** and **500T** discontinued.

1955

Manx: no change.
Inter: welded subframe, full width hubs.
ES2: light alloy head, Monobloc, new seat, new bars, horn button built in on right, boxed in rear number plate with rear reflector, new front number plate, smaller tank, thinner kneegrips.
19S, **19R**: added to range with 597 cc engines as **ES2**. Rigid model with 4·00 × 18 rear tyre.
Late in year, **19R** discontinued.

1956

Manx: all 1955 works changes with hollow exhaust valve head, flanged double row ballrace main on timing side, full ring clutch discs.
Inter: not listed but built to special order in very small numbers.
50: introduced with alloy head, 7·3:1 compression ratio.
ES2: raised compression ratio to 7·1:1, improved exhaust cam, more cylinder fin area, larger exhaust pipe.
19S: cam and fins as **ES2**, ratio to 6·4, same pipe.
All ohv: full width hubs front and rear, larger petrol tank with screw held kneegrips, new taps and fuel lines, new oil tank styled to match combined toolbox and battery carrier, regulator in toolbox, deeper headlamp shell with speedometer, ammeter and switch in panel, pilot light in main reflector, combined horn push and dipswitch on left bar, larger mudguards, softer front suspension, improved propstand, altered oil lines, speedo drive from below wheel spindle, forged end lugs for rear fork, flat-top dualseat, Armstrong rear units, during year AMC gearbox adopted.

1957

Manx: sleeved big-end, modified rod and crankpin, sodium cooled inlet, bigger carb, coarse pitch bevel, new cam for 350.
Inter: tubular silencer, altered seat, separate chrome tank panels, built to special order in very small numbers, often using **Manx** parts modified to suit.
All ohv: new head with integral pushrod tunnels and inlet flange, cams, frame all tubular and no cradle casting, narrow rear fork pivot, no rear brake torque arm, Girlings, small trail increase, new front hub and brake backplate with forged cam lever, end cover for mag-dyno, enclosed rear brake pedal spring, modified rear chainguard, no panel in headlamp shell, plastic bar grips, tubular silencer with no tailpipe, tank with separate side panels, optional air filter.

1958

Manx: AMC box.
Inter: last few built as in 1957
All ohv: no change.
Late in year **19S** discontinued.

1959

Manx: needle roller vertical shaft with splined drive so no bottom bevel housing, camshaft bearings altered, no sodium cooled valves, Nimonic exhaust, clutch inserts bonded to driven plates, one plate less.

50, ES2: AC electrics, coil ignition, coil under tank rear, rectifier under tooltray under seat, points with auto-advance in housing in old magneto position, chaincase altered, covers over engine plates, multi-rate valve springs, featherbed frame, rear chaincase option, sidecar trail and suspension option, 3·5 gallon tank.

1960
Manx: wider rod eye, eccentric big-end thrust washers, cr 11:1, chrome top ring, non-rotating circlips, magnetic drain plug, cap screws for timing side main, tappets with Stellite tips, deeper top housing spigot, damped weir float, reinforced clutch centre, revised gear pedal linkage, fluted megaphone flat, stronger rear sprocket dowels, serrated rims, oil tank strap, glass fibre seat base and front number plate, perspex flyscreen, 9000 rpm counter.
50, ES2: gearbox internals and gearing altered.

1961
Manx: single Dykes ring piston, shorter barrel, GP2, altered valve gear, clip and stop for megaphone, oil tank mounted on rubber pivots and held by bands, tubular rear fork pivot.
50, ES2: slimline frame, narrower tank, new qd seat.

1962
Manx: machines built from spares with dual 7 in. 2ls front brake, bigger capacity oil pump, larger oil feed through crankshaft, barrels machined at both ends to avoid thinning top fin, 350 carburettor larger, 500 carburettor spacer with taper bore, 500 with more tank cutaway at rear to clear bellmouth.
50, ES2: no change, chaincase option continued.

1963
50, ES2: no change, discontinued late in year.

1965
50 Mk 2, ES2 Mk 2: AMC model with Roadholder forks and Norton wheels.

1966
50 Mk 2, ES2 Mk 2: no changes, discontinued late in year.

Norzi or Guzton special; so many featherbed frames have housed 'foreign' engines

Model charts

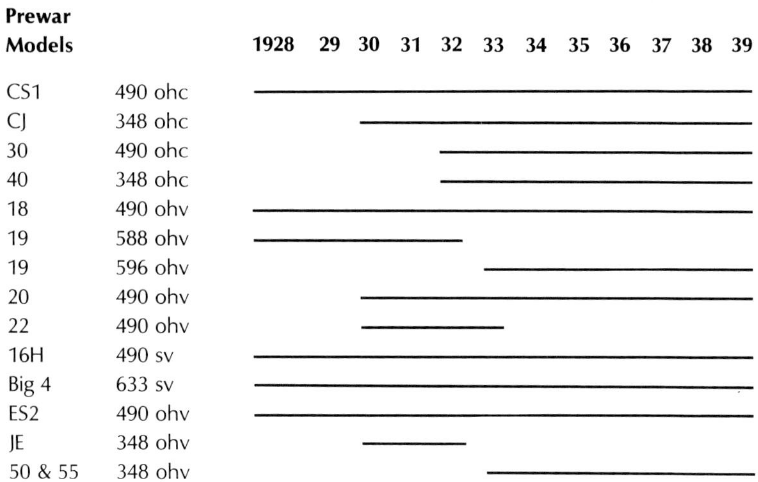

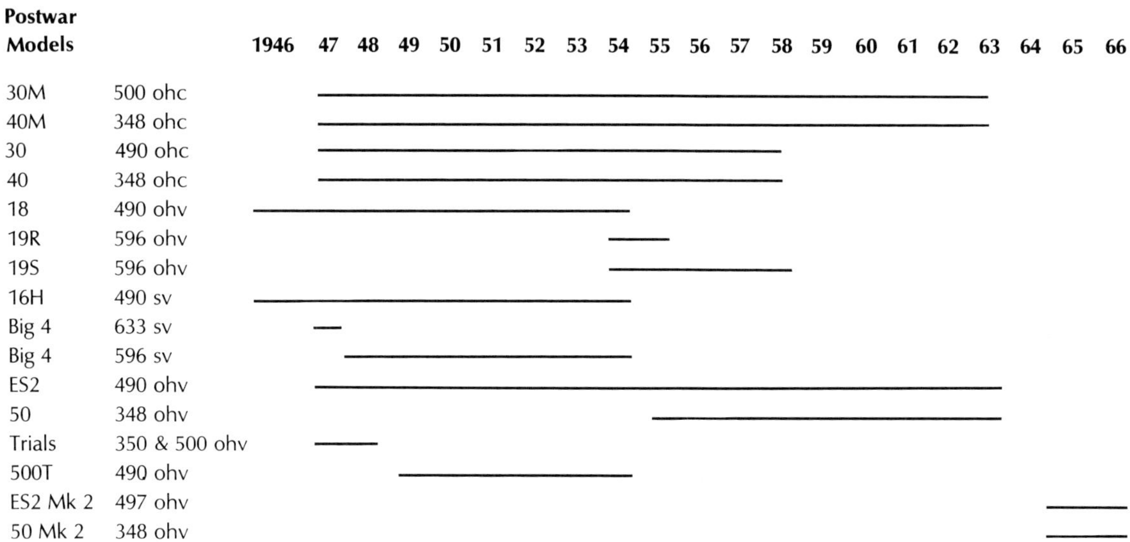

Nannette Bowers-Stel

Goldsmith and Jeweller

Cv / Vi s ua l Ar t s Re s e a r c h Series 5 8

NANETT E BOWERS - ST EL
Go l dsmi th

ISBN 19781904727538
ISSN 1476-9980

This interview was first published in
Cv Journal of the Arts
(London: The Artist's Journal Ltd 1989)
British Library Cataloguing in Publication Data.
A catalogue record of this book is available from the British Library

A digital format publication from
Cv Publications,
10 Barley Mow Passage, London W4 4PH
www.tracksdirectory.ision.co.uk

Cv/Visual Arts Research

The survey began in April 1988 as interviews with jewellers, fashion designers and furniture restorers, based at Old Loom House in Whitechapel, launching a quarterly review *Cv Journal of Art and Crafts*. *Cv Journal* was published to 1992 and the collection of interviews, features and reviews provided the foundation of the Cv/Visual Arts Research archive and subsequent publications. Following *Cv Journal* the data-base shifted towards digital publishing, allowing a greater flexibility of communication.

Cv/*VAR* addresses the fields of academic research, galleries and museums worldwide, and a non-specialist readership. In this respect the archive has been re-organised as a file system which may be accessed as individual articles or collated volumes, according to specific requirements. The data base is categorized as *Interviews-Artists; Curators and Collections; CraftsDirectory; Small Histories; Social Studies and Studio Work.* Titles are published as conventional books and information files made by digital process as print on demand, as well as CDs and DVDs in Cv Publications' software catalogue. There are now over ninety titles published in the series.

Cv/VAR series no.58 features an interview by Sarah Batiste recorded in 1989 with dutch-born goldsmith and jeweller Nannette Bowers- Stel, which explores the development of her practice, giving tips on equipment and client relations to starter designers.

Contents

Nannette Bowers-Stel

Interview by Sarah Batiste recorded 9.11.88

SB How long have you been working as a jeweller?

NBS. About twelve years

What started your interest in jewellery?

I was sent on a course when I was ten years old. In one classroom they did metalwork. I was always more interested in metalwork, so I started from there and it all happened.

Did you go to a college of art?

No, I was an apprentice for three and a half years in Holland. And after that I worked for a year for a silversmith.

That gave you a good grounding?

It did, yes.

NANNETTE BOWERS STEL

GOLDSMITH & JEWELLER

Is that the usual route for a jeweller to follow, to do an apprenticeship?

It was but it's dying out now. People now go to college; in my country it was a specialised college in Schoonhoven in Holland.

How long have you been over in England?

Eight years.

What brought you here?

A job. I had done my apprenticeship and worked for the silversmith. Then they advised me to go to Germany or England, so I chose England. I had a job for three years with Charles de Temple in Jermyn Street, and I learnt a lot there. They specialise in hand-made jewellery.

Classical style?

No, very modern designs.

Are there designs you're particularly fond of?

NANETTE BOWERS STEL

GOLDSMITH & JEWELLER

None in particular. I follow what's going on, what's new at the exhibitions and what's happening, but you have to be careful that you don't take over someone else's style; keep to your own style.

What about metals, what metals do you enjoy working with?

Gold and silver, precious metals.

And gems?

I work a lot with gems, and plain jewellery. It depends.

Where do you sell?

At the moment I sell a lot in Holland and I sell in Norfolk, to some little craft galleries.

Well established firms?

Yes.

Looking at samples

NANNETTE BOWERS STEL

GOLDSMITH & JEWELLER

It's a hatpin?

No, actually it's a stickpin, that you have on a coat or a sweater.

It's got a fan shape at the top which is very beautiful, and lines?

They are engraved lines and I like the contrast between the matt and highly polished surface.

We've got a necklace here with a triangular shape - is it silver? And what is the black material?

Ebony, with a snake chain. It is very flexible so it lies very well around the neck. I think it is a lovely combination. I have the ebony waxed, not polished.

What is in demand?

Earrings are very popular, but when you go into shops they say, can you make something else, because everyone is making earrings. So I do

NANETTE BOWERS STEL

GOLDSMITH & JEWELLER

bracelets, chains, pendants as well.

This piece has a moon with a star attached to it. Is the star made of diamonds?

Yes, all diamonds, it was quite spectacular. The moon was taken from the client's Coat of Arms, it has a special meaning to it.

Do people come with a specific idea of what they want?

Mostly, but some say, I haven't a clue, so I make some drawings and we go from there.

Do you ever get commissions you don't necessarily like doing?

Sometimes, and it's hard if it's not to your own taste. Still, if that's what the customer wants, that's fine.

How hard is it for somebody just out of college to set up on their own? I mean you've got a daunting amount of equipment here.

NANNETTE BOWERS STEL

GOLDSMITH & JEWELLER

I suppose it is hard, but all the schemes for new businesses must help, and the Crafts Council does a lot to help people. But it takes a while to get established and you have to keep the standard of your work very high to compete. Not only with individual jewellers, but the mass production companies as well. People see those prices in the shops and then see the difference when they come to see you for handmade work. I often find the shops aren't interested to take on commissions because they're too time consuming.

To set up as a jeweller, presumably the equipment could be bought second-hand?

It can. I built mine up very slowly over twelve years. First when I was an apprentice and I got tools from the workshop; then I won some prize

NANETTE BOWERS STEL

GOLDSMITH & JEWELLER

money with which I bought my first pair of pliers and one hammer. The tools can be built up bit by bit, but the big costs are the polishing machines, extractor fans, casting machines, and a rolling mill to make metal thinner. Those are expensive.

Biographical Note

Nannette Bowers-Stel was born in 1950 in Rotterdam, Holland. From 1962-68 she was educated at the HBS School in Groningen, where her family moved in 1951. She studied metalwork at college in Schoonhoven and obtained a City and Guilds Certificate in England in 1984. She was employed in Holland by J.Venema, 1976-79, and C.Wolf, 1979-80. She then worked for Charles de Temple in Jermyn Street, London, from 1980-83. She exhibited her work with Maya Wildevuur, Paterswolde Hooghalen in 1988, and Alby Crafts, Norfolk. She established her practice in London in 1980, and married in England in 1981.

The Researcher: Sarah Batiste studied fine art at Goldsmiths and Wimbledon School of Art.

NANNETTE BOWERS STEL
GOLDSMITH & JEWELLER

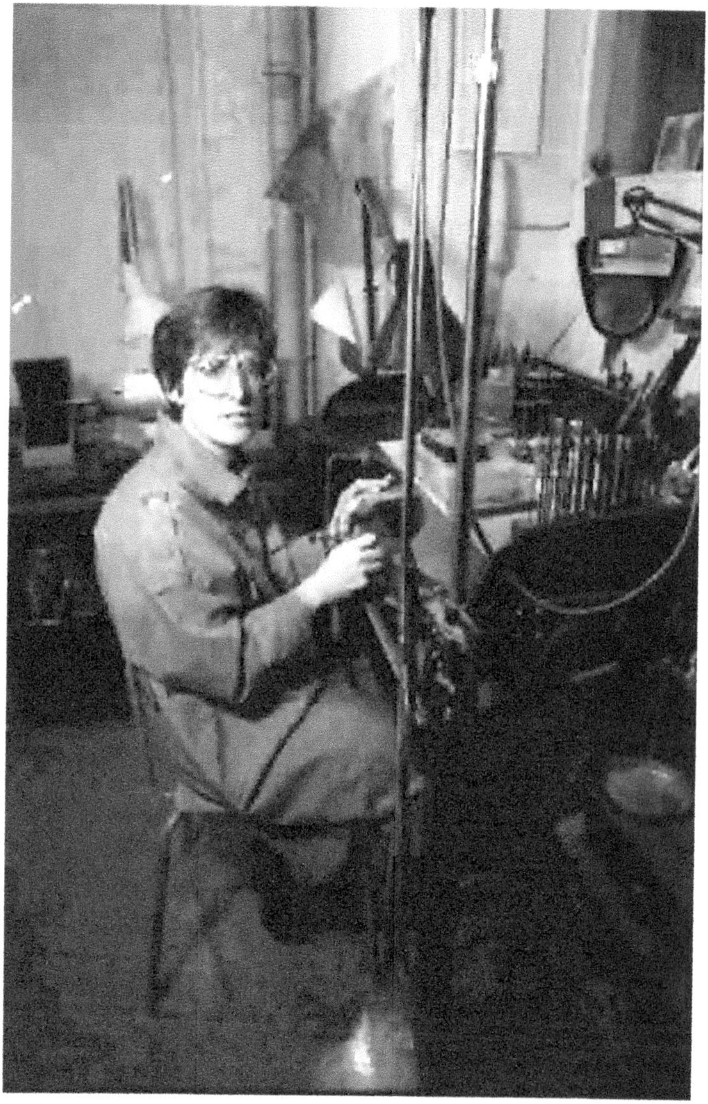

Nanette Bowers-Stel in her workshop
at The Barley Mow Centre, Chiswick, 1989 Cv/VAR Archive

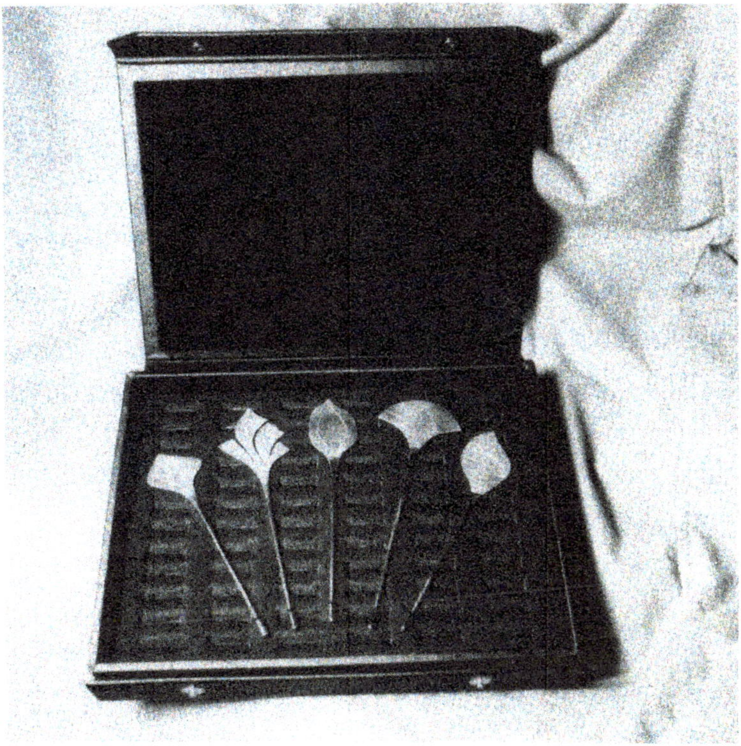

Nannette Bowers-Stel : Stickpins
Cv/VAR Archive

ISSN 0954/1608

Cv

Journal of Art and Crafts

Second Issue £1.50
February 1989
Volume 2 Number 1

Nannette Bowers-Stel

Wilbraham and Vaughan

Georgina Hadrill

Anthea Lahr

Sara Hayward

Guillem Ramos-Poquí

Sandra Smith

Marika

Special Events

Book Choice

Style by Stylists/Counter Check/Chain Link

Cv Journal of Art and Crafts Volume 2/1 February 1989 Cv/VAR Archive

Cv Journal of Art 2/1 Frontispiece Guillem Ramos-Poqui, Sara Hayward, Marika Wenman, Georgina Hadrill Cv/VAR Archive

Nanette Bowers-Stel
Goldsmith

Nannette Bowers-Stel *Interview by SB 9.11.88*

How long have you been working as a jeweller?

About twelve years.

What started your interest in jewellery?

I was sent to a course when I was ten years old. In one class you did paperwork, and in the next classroom they did metalwork. I was always more interested in metalwork, so I started from there and it all happened.

Did you go to a college of art?

No, I was an apprentice for three and a half years in Holland. And after that I worked for a year for a Silversmith.

That gave you a good basic grounding.

It did, yes.

Is that the usual route for a jeweller to follow - to do an apprenticeship?

It was, but it's dying out now. People now go to college. A specialised college in Schoonhoven in Holland.

Are there many jewellers working there that you're familiar with? That you still perhaps have contact with?

Yes, but I've got to know people here as well.

How long have you been over in England?

Eight years.

What brought you here?

A job. I had done my apprenticeship and worked for the Silversmith. Then they advised me to go to Germany or England, so I chose England. I had a job for three years at **Charles De Temple** in Jermyn Street, and that was very nice, I learned a lot there. They specialise in hand-made jewellery.

Classical?

No, modern, very modern designs.

Did that inspire you?

Yes.

Are there designs you're particularly fond of?

Not one particularly. I follow what's going on, what's new at the exhibitions, and what's happening. But you have to be careful that you don't take over someone else's style—keep to your own style.

What about metals? What metals do you enjoy working with?

Gold and silver. Precious metals.

What about gems?

I work a lot with gems, and just plain jewellery. It depends.

And finding outlets for your work, where do you sell?

At the moment I sell a lot in Holland, and I sell some in Norfolk, to some of the little craft galleries.

Well established firms?

Yes. *(looking at samples)*

It's a hatpin.

No, it's actually a stickpin, that you have on a coat or a sweater.

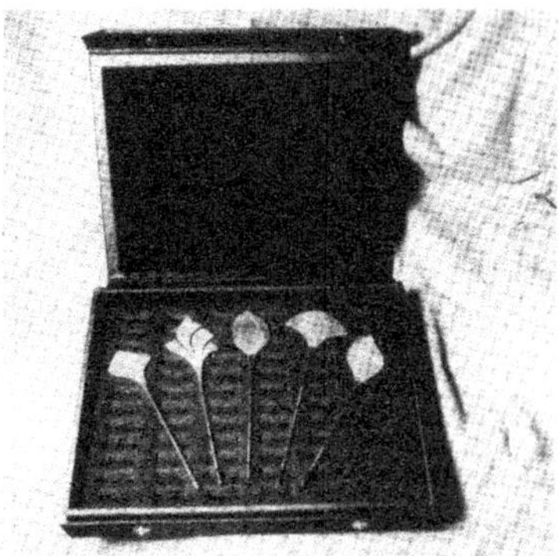

Stickpins by Nannette Bowers Stel 1988

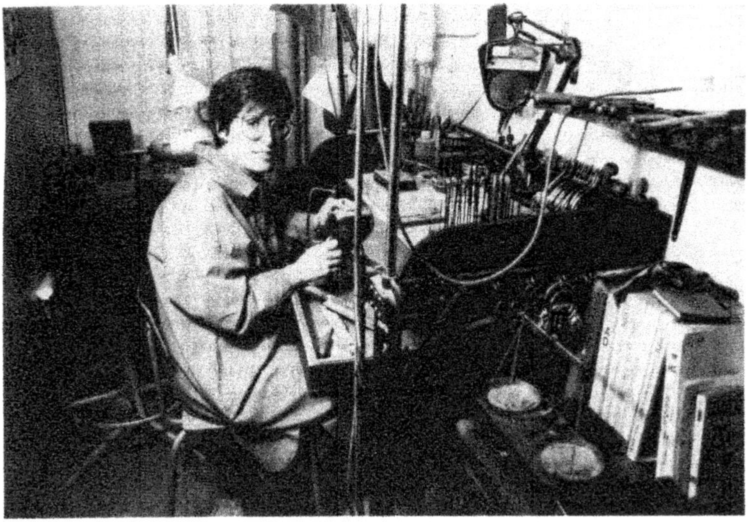

It's got a fan shape at the top which is very beautiful, and lines.

They are engraved lines, and I like the contrast between the matt and the highly polished surface.

CHAIN LINK

We've got a knecklace here with a triangular shape - is it silver? And what is the black material?

Ebony, with a snake chain. It is very flexible so it lies very well around the neck. I think it is a lovely combination. I have the ebony waxed not polished.

That's a little brooch.

It is, yes.

..With shiney details in gold on the bobble, on matt silver. Very pretty.

Yes I love that.

Cufflinks and more earrings, very nice indeed. Do you work with materials apart from ebony?

No. It's ebony or plain silver, or gold.

Ebony shows up silver very well.

I think so, yes. At the moment I'm experimenting with rosewood and silver, that's the next step.

With what designs, similar to these?

Similar, but not repeating the shapes..of little half moons, and diamond shapes.

Do you have much call for a particular piece of jewellery, like knecklaces.

Earrings. Earrings are very popular. But when you go to the shops they say, can you make something else because everyone is making earrings. So I do bracelets, chains, pendants as well.

This piece here looks like a shell, is it a brooch?

It isn't. It was just an ornament made as a commission. The person had the shell and he wanted it made in silver.

Beautiful. That detail—is it a flower on top?

Yes, but that was really how the shell was.

It's almost like a fossil on top of it.

It had a sentimental value for him, the shell.

Do you look at nature a lot?

I do, yes.

Because all your designs have an organic feel to them, the sort of things which one might find in rocks. It has that feel to it.

Yes I like that.

This ring, is that gem an emerald.

Yes, it was a commission from a person who had bought the stone abroad, then came to me, and wanted a ring made in gold. This one is a brooch with a tiny diamond in it.

It almost looks like something you might find in the sea.

These earrings are fan shaped and so it completely depends on what the person wants—what carat gold they want, how big they want it. I price it individually, according to each commission. In 18 carat gold, about £300.

You don't do the enamelling yourself.

No, not the letters, that's a specialist's job.

This one has a moon with a star attached to it. Is the star made of diamonds?

Yes, all diamonds. It was quite spectacular. The moon was taken from their Coat of Arms, it had a special meaning to it. There was a silver moon with a gold setting and the diamonds.

How do you find your commissions?

People tell other people.

Word of mouth?

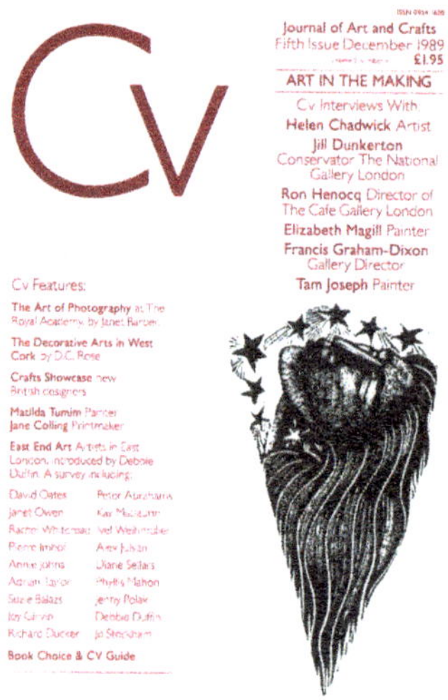

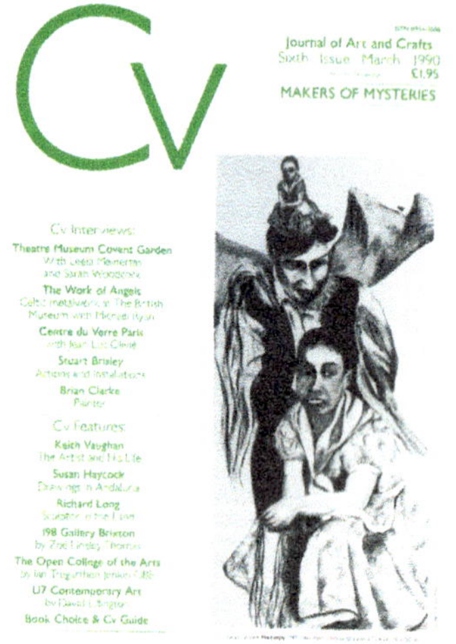

Cv Journal covers V 1/1- 3/4 1988-90 Cv/VAR Archive

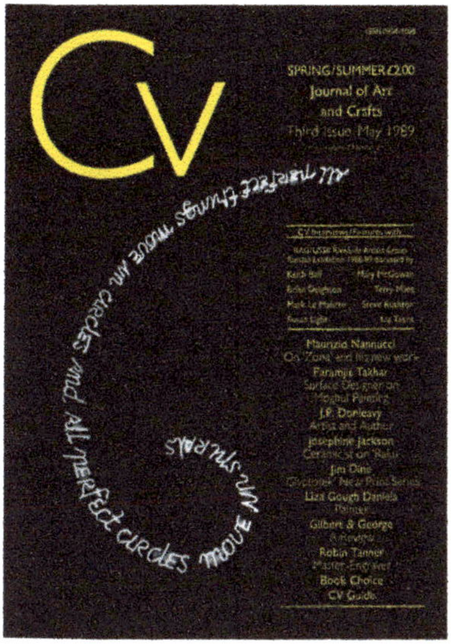
Cv
SPRING/SUMMER £2.00
Journal of Art
and Crafts
Third Issue May 1989
All imperfect things move in circles and all perfect circles move in spirals
Maurizio Nannucci
Paramjit Takhar
J.P. Donleavy
Artist and Author
Josephine Jackson
Jim Dine
Liza Gough Daniels
Gilbert & George
Robin Tanner
Book Choice
CV Guide

Cv
Journal of Art and Crafts
£2.00
SCULPTURE AND
PAINTING NOW
John Skelton MBE
Richard Wilson
Judy Brown
Hughie O'Donoghue
ANDY WARHOL: A RETROSPECTIVE

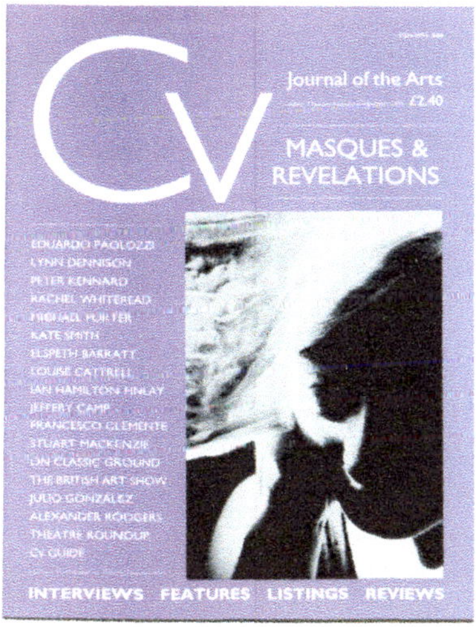
Cv
Journal of the Arts
£2.40
MASQUES &
REVELATIONS
INTERVIEWS FEATURES LISTINGS REVIEWS

Cv
JOURNAL of the ARTS
December – February 1991
JOHN GOTO
JAKE TILSON
JUNE REDFERN
AMIKAM TOREN
HISTORY MEMORY DESIRE

Art . Crafts . Careers. Travel .
Social Studies & Studio Work
Published by Cv Publications
www.tracksdirectory.ision.co.uk